Nelson Maths

1

Pupil Book

Karen Morrison
Lisa Greenstein

OXFORD
UNIVERSITY PRESS

OXFORD
UNIVERSITY PRESS

Great Clarendon Street, Oxford, OX2 6DP, United Kingdom

Oxford University Press is a department of the University of Oxford.

It furthers the University's objective of excellence in research, scholarship, and education by publishing worldwide. Oxford is a registered trade mark of Oxford University Press in the UK and in certain other countries.

British Library Cataloguing in Publication Data

Data available

ISBN: 978-1-382-00998-0

1 3 5 7 9 10 8 6 4 2

Paper used in the production of this book is a natural, recyclable product made from wood grown in sustainable forests. The manufacturing process conforms to the environmental regulations of the country of origin.

Printed in Great Britain by Bell and Bain Ltd, Glasgow

Acknowledgements

The publisher and authors would like to thank the following for permission to use photographs and other copyright material:

Cover: Matthieu Nivesse. **Photos: p36(a):** Broeb/Shutterstock; **p36(b):** Vichy Deal/ Shutterstock; **p36(c):** Arthito/Shutterstock; **p36(d):** Brocreative/Shutterstock; **p36(e):** litchima/ Shutterstock; **p36(f):** cigdem/Shutterstock.

Artwork by Aviel Basil, Bernard Adnet, Andy Elkerton, Kathryn Mitter, David Lopez, Steve Cox/ PFD, Integra Software Services, Alan Rogers, Pantek Media, Maidstone, and OKS Prepress India.

Every effort has been made to contact copyright holders of material reproduced in this book. Any omissions will be rectified in subsequent printings if notice is given to the publisher.

Contents

Unit 1 Think maths   **5**
Say what you can see 5
Connect the dots..6
How we learn maths 7

Unit 2 Measure and compare   **8**
Compare length...8
Measure length..9
Full or empty 10
Which holds more? 11
Solve capacity problems........................12
Measuring instruments 13

Unit 3 Count to 10 and beyond   **14**
Numbers to 10...14
Zero to 10..15
Number names..16
1 more, 1 less...17
Even and odd numbers...........................18

Unit 4 2D shapes   **19**
Shape names...19
Shapes and patterns...............................20
Corners and sides.....................................21
Sorting shapes 22
Which doesn't match?.............................23

Unit 5 Order and position   **24**
First, second, third... 24
Numbers in order......................................25
Left and right, up and down 26
On and under, in front and behind 27

Unit 6 Count to 20   **28**
Numbers to 20..28
Count shapes 29
Count in 2s..30
Count in 5s..31
Tens and ones 32

Unit 7 Mass   **33**
Heavier 33
Lighter 34
Talk about mass 35
Weighing 36
How much does it weigh? 37
Solve mass problems 38

Mixed practice 1   **39**

Unit 8 Add and take away   **41**
Add to make 10..41
Count on..43
Add..44
Totals..45
Count back...46
Take away 47
Add and take away....................................48

Unit 9 Time   **49**
Times of day...49
Days of the week.......................................50
Quicker and slower...................................51
Talking about time....................................52
Hours, minutes and seconds................54

**Unit 10 More adding and
taking away**   **55**
Make 10 55
Fact families 56
Make 20..57
What's missing? 58
Find the difference...................................59

Unit 11 Capacity and temperature   **60**
Full, half full or empty 60
Which holds more? 61
Estimate and measure 62
Hot or cold 63
How warm or cool is it? 64

Unit 12 3D shapes   **65**
Name 3D shapes..65
Describe 3D shape 66
Alike and different 67
Which shape doesn't match?.................68

Unit 13 Multiply   **69**
Make rows and columns.........................69
Skip count 70
Multiply..71
Doubles...72
Arrays..73
Different ways to multiply 74

Unit 14 Zero to 100 — 75
Bigger numbers 75
Zero 76
Number lines and charts......................... 77
Count on a number chart........................ 79
Rocket countdown.................................. 80
Fast track, slow track............................. 81

Mixed practice 2 — 82

Unit 15 Position and direction — 84
Position words... 84
Directions... 85
Forwards and backwards 86
Half turns and full turns 87
Left and right 88
Part of a turn... 90

Unit 16 Money — 92
Coins in my country............................... 92
Recognise coins...................................... 93
Make totals 94

Unit 17 Sort shapes — 95
Properties of shapes............................... 95
More properties of shapes 97
Sort 2D shapes....................................... 98
Sort 3D shapes....................................... 99
More sorting 100

Unit 18 Share — 101
Halves 101
Make halves.. 102
Half of a number 103
Divide 104
Quarters 106
Quarter of a number 107

Unit 19 More about time — 108
What is the time?.................................. 108
Different times of day 109
Half past 110
My daily routine 112
Use time words...................................... 114

Unit 20 Data — 115
Use a Carroll diagram 115
Venn diagrams...................................... 116
Pictograms.. 117
More pictograms 118
A cap pictogram 119
Block diagrams 120

Mixed practice 3 — 121

Glossary — 124

Think maths

Say what you can see

> 💭 **Think and share**
>
> Look at these **patterns**.
>
> What is the same and what is different?
>
>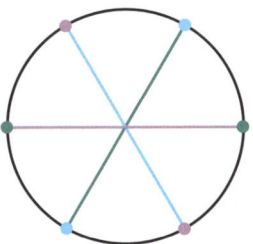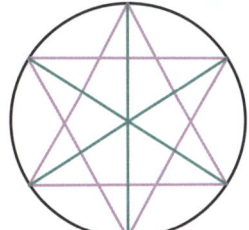

1 Copy or trace this circle. Use parts of circles and lines to make patterns like the ones above.

2 Kayla and Sai tried to make patterns like the ones above. What mistakes have they made?

Kayla   Sai

Connect the dots

We can connect things in many ways.

We all make connections in different ways and we all learn in different ways.

1 **a** Draw 5 dots.

b Connect the dots.

c Work with a partner and compare your pictures.
What is the same and what is different?

d **Add** 3 more dots to your picture.

e Draw more lines to connect the dots.

2 Compare pictures with a friend.
What pictures did you make?
How did you make them?

How we learn maths

We can work together or alone.

We can work slowly or fast.

We can ask questions and help each other.

1 Match words to the pictures.

A

B

C

D

talk	draw	build
help	explain	ask
share	think	use

➡ *Workbook page 4*

Measure and compare

Compare length

> 💭 **Think and share**
>
> Why are these things long?
>
>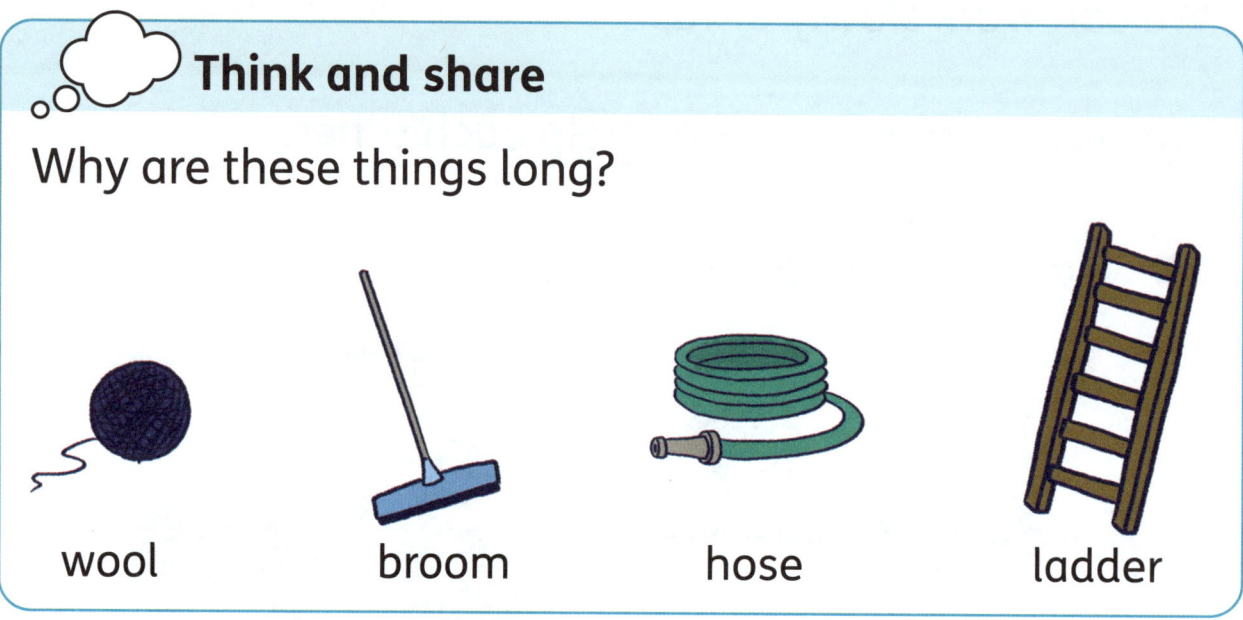
>
> wool   broom   hose   ladder

1 Which is **taller** in each pair?

a

b

2 Which is **shorter** in each pair?

a

b

➡ *Workbook page 5, page 6 and page 7*

Measure length

Length means how long something is.
We can use different things to **measure** length.

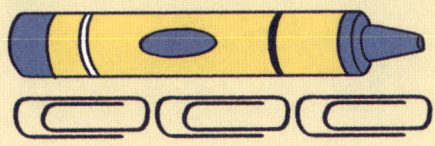

1 About how many 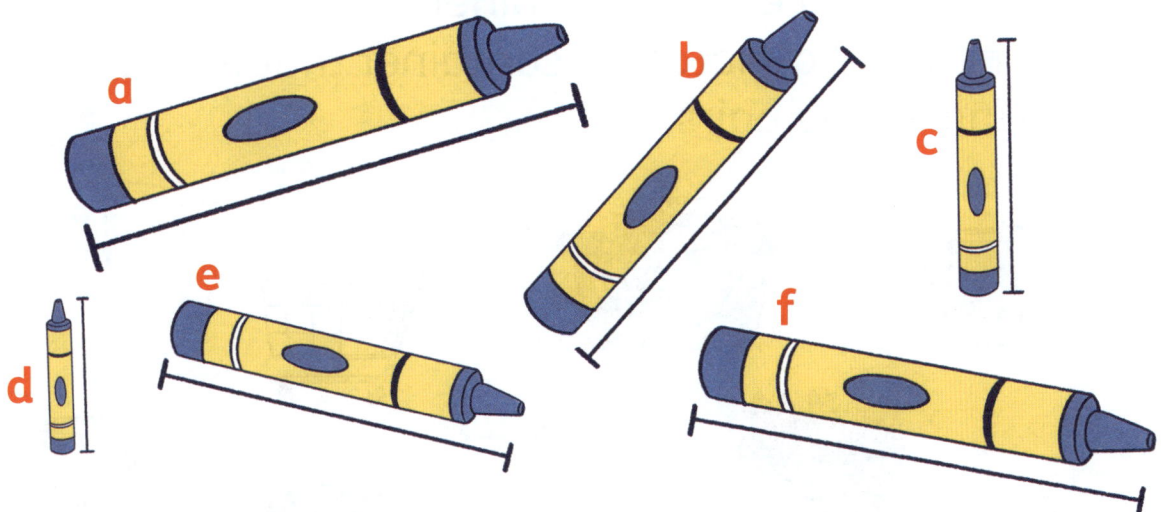 long is each crayon?
Use real paperclips to measure.

a

b

c

e

d

f

2 This is a ruler.

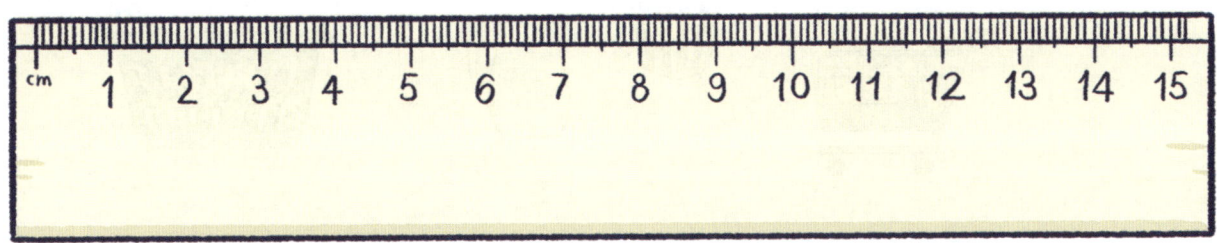

How can you use a ruler to measure the crayons?

➡ *Workbook page 8*

Full or empty

full　　**empty**

1 Point to an empty container.
Then find a matching container that is full.
Find all the pairs.

2 What do you see around your home or
classroom that is empty?
What do you see that is full?

➡ *Workbook page 9*

Which holds more?

The **bigger** bottle holds more than the **smaller** one.

The **wider** bucket holds more than the **narrower** one.

1 Which one in each pair holds more?

a

b

c

d

Solve capacity problems

Remember that **+** means add or count on.
= means makes or is **equal** to.

 **Problem solving**

3 cups fill 1 bottle.

3 bottles fill 1 bucket.

1 How many cups fill one bucket?

2 Work with a partner.
Make up a question like the one above and ask your partner to solve it.

Draw a picture to help you.

Measuring instruments

Measuring instruments are special tools we use to measure.

1 What do we use it for?

a

tape measure

b

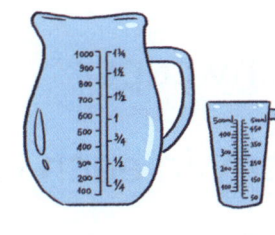

measuring jugs

c

measuring spoons

d

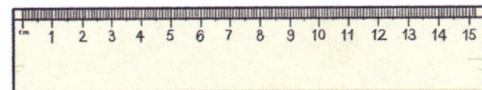

ruler

2 How can you measure your height?

Count to 10 and beyond

Numbers to 10

Think and share

Count.

How many cups?

How many in each row?

1 Count the number of each type of fruit.

➡ Workbook page 10 and page 11

Zero to 10

3 dots   3 cherries

1 Count how many in each set. Find a matching set of dots. Count with a partner.

➡ *Workbook page 12*

Number names

1 Read each number name.
Find a set with the same number.

eleven   twelve   thirteen   fourteen

fifteen   sixteen   seventeen   eighteen

nineteen   twenty

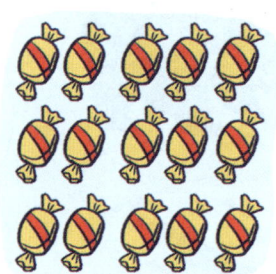

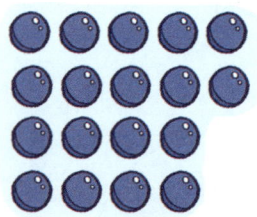

➡ *Workbook page 13*

1 more, 1 less

2 frogs on the leaf.
1 more.
Now there are 3.

4 birds on the branch.
1 less.
Now there are 3.

1 Say how many.

a 2 playing.
1 more joins.
How many?

b 3 coconuts on the ground.
1 more falls.
How many?

c 5 birds.
1 less.
How many?

d 6 pencils.
1 less.
How many?

➡ *Workbook page 14 and page 15*

Even and odd numbers

How does the **pattern** grow?

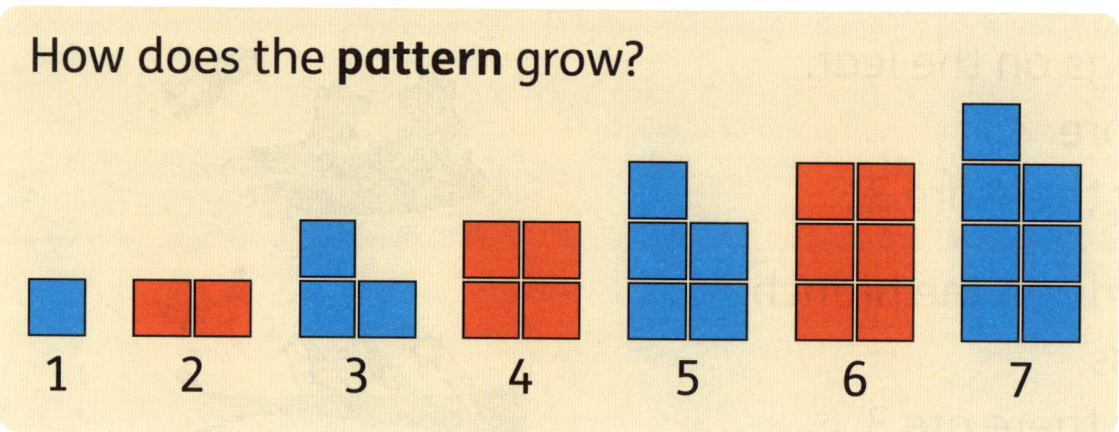

| 1 | 2 | 3 | 4 | 5 | 6 | 7 |

1 Count in 2s, like this: 2, 4, 6, . . .
What do we call these numbers?

1	2	3	4	5	6	7	8	9	10
11	12	13	14	15	16	17	18	19	20

Problem solving

2 Which shapes have an **odd number** of blocks?

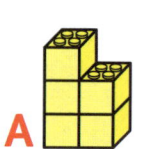 A
 B
 C
 D
 E
 F

3 **a** What kind of number do you get if you add
1 to an **even number**?

b What kind of number
do you get if you add 1
to an odd number?

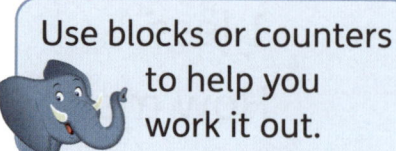

Use blocks or counters
to help you
work it out.

➡ *Workbook page 16*

Shape names

 **Think and share**

These are **2D shapes**.

 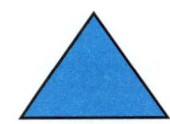  

circle   **triangle**   **square**   **rectangle**

Which shape looks very different from the others? Why?

1 Read the shape names.
Count how many of each shape.

a

circles

b
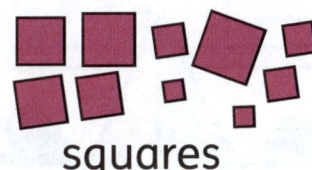
squares

c

triangles

d

rectangles

 **Problem solving**

2 Put squares together to make
new squares and rectangles.
What do you notice?

 Which takes more squares to make?

➡ *Workbook page 17*

Shapes and patterns

We can see shapes all around us.

Sometimes they make patterns.

1 Find triangles, circles, squares and rectangles in the picture.
Which shapes repeat to make a pattern?

2 **a** Find something in the classroom that is shaped like a rectangle.

b What things in a kitchen are shaped like a circle?

➡ *Workbook page 18*

Corners and sides

When you describe a shape, you say how it looks.

Triangles, squares and rectangles have **sides**. Each side is a straight line. Sides meet at **corners**.

A circle has no corners and no sides. The outline of a circle is curved.

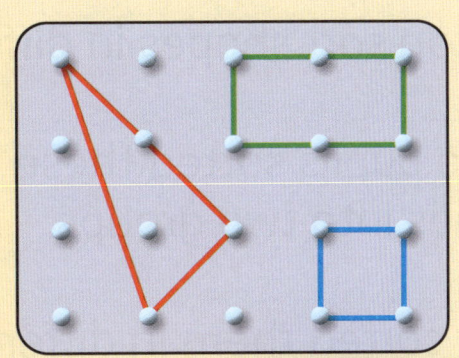

1 Describe each shape. Which ones are the same shapes?

a

b

c

d

e

f

g

h

➡ Workbook page 19

Sorting shapes

We can sort shapes in different ways.

Some pupils sorted these shapes into two groups.

They sorted them in different ways.

by colour

by shape

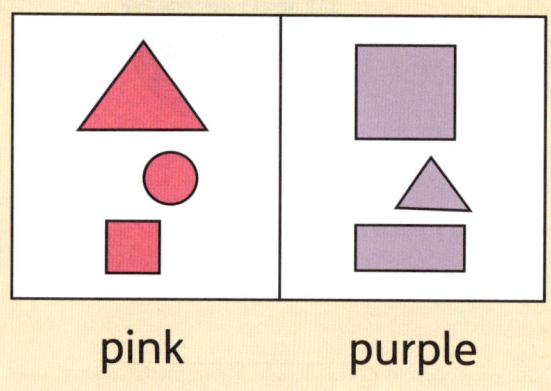

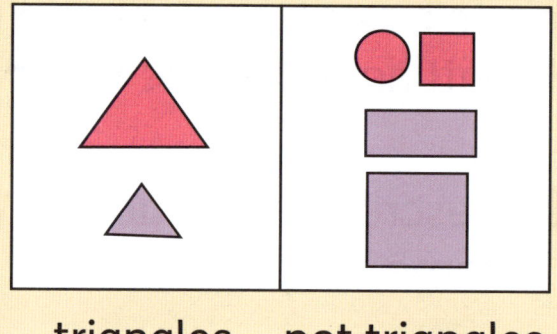

pink   purple   triangles   not triangles

1 How was each group sorted?

a

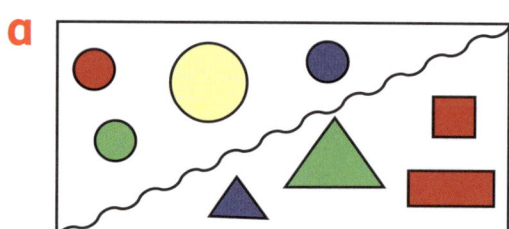

b

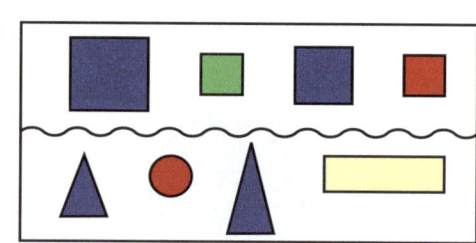

c

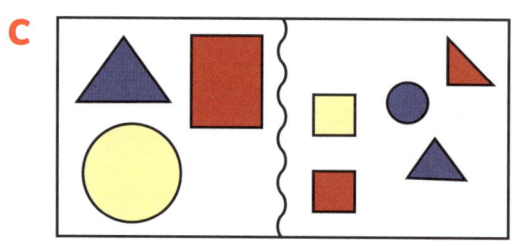

d

Which doesn't match?

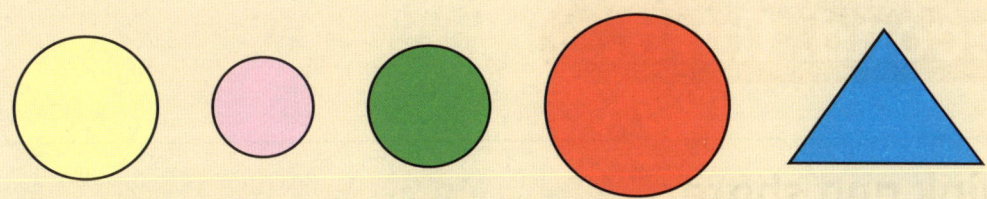

The triangle doesn't match.

The other shapes are all circles.

1 Which one doesn't match in each set?
Say why.

➡ *Workbook page 20*

Order and position

First, second, third...

> 💭 **Think and share**
>
> The **first** scoop of ice cream put on the cone is pink.
>
> Which colour is **second**?
>
> Which is **third**?

 ← first

1 Look at the 10 aliens.
The first alien is sitting.
Say what each alien is doing.
Use the words below.

1st first	2nd second	3rd third	4th fourth	5th fifth
6th sixth	7th seventh	8th eighth	9th ninth	10th tenth

➡ *Workbook page 21 and page 22*

Numbers in order

The numbers are mixed up.

You can put them in the correct order.

1 These numbers are mixed up.
Say the correct order.

a  7   5   6

b

c 2   4   3

d  10   12   11

e

➡ *Workbook page 23 and page 24*

Left and right, up and down

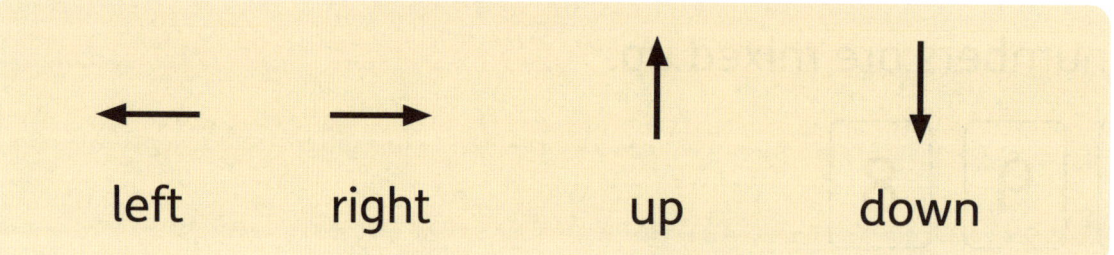

left   right   up   down

 **Problem solving**

 When you give directions, you say which way to go.

1 Give directions to get the mouse to each block of cheese and then to her house.

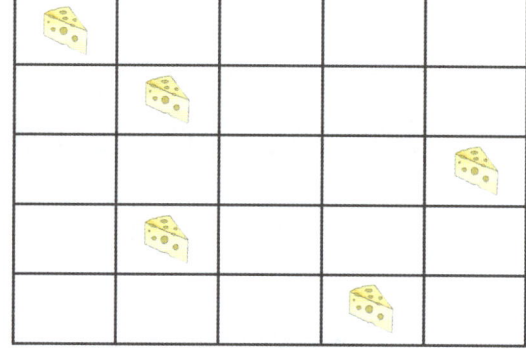

2 Use this grid and 5 counters. Pretend the counters are pieces of cheese. Make your own direction puzzle. Give it to a partner to solve.

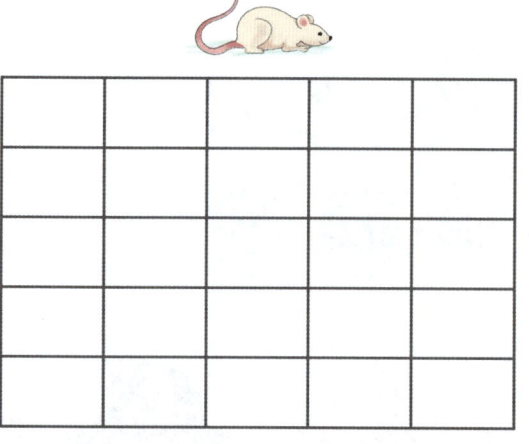

On and under, in front and behind

The cat is on the chair. The pillow is under the cat.

The bucket is behind the sandcastle. The spade is in front of the sandcastle.

1 Where is the cat?

a  b  c

2 Where is the ball?

a  b  c

UNIT 6 Count to 20

Numbers to 20

💭 **Think and share**

Count. What patterns can you see?

1	2	3	4	5	6	7	8	9	10
11	12	13	14	15	16	17	18	19	20

1 Count.

a  b  c

d  e 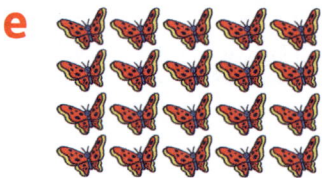 f

💡 **Problem solving**

2 What pattern do you see in the groups of bugs?

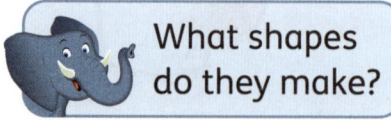

What shapes do they make?

➡ *Workbook page 25*

Count shapes

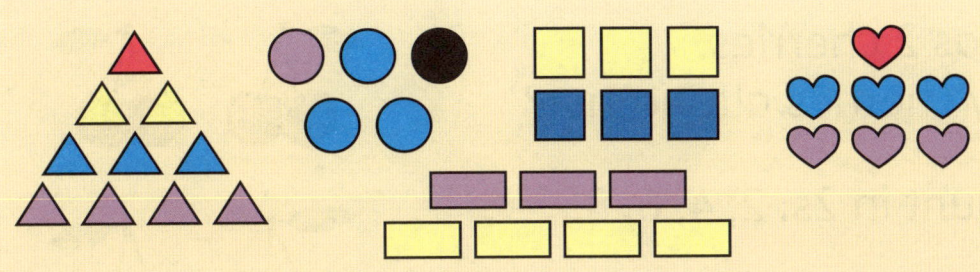

How many △ + ☐ ?

10 triangles + 6 squares. 16 altogether.

1 Use the shapes above to work out how many altogether.

a ◯ + ☐

b △ + ◯

c △ + ▭

d ♡ + △

e △ + ▭ + black shapes

f purple shapes

 **Problem solving**

2 Can you make groups of shapes for all the numbers from 1 to 20, using shapes in the picture?

➡ *Workbook page 26*

Count in 2s

A bunch has 2 cherries.
How many cherries altogether?

We can count in 2s: 2, 4, 6, 8.

There are 8 cherries.

1 Count in 2s and say how many.

a A stem has 2 flowers.
How many flowers altogether?

b A bug has 2 spots.
How many spots altogether?

c A button has 2 holes.
How many holes altogether?

d A pot has 2 handles.
How many handles altogether?

2 Count in 2s to 20.

1	2	3	4	5	6	7	8	9	10
11	12	13	14	15	16	17	18	19	20

➡ *Workbook page 27*

Count in 5s

We can count forwards or backwards.

We can count in 5s.

1 Count down in 1s from 20 to 0 for the rocket to blast off.

20 **15** **10** **5**

19	14	9	4
18	13	8	3
17	12	7	2
16	11	6	1
			0

BLAST OFF!

2 Count forwards and then backwards in 5s. Use the numbers with stars around them.

Problem solving

3 What number patterns do you notice?

➡ *Workbook page 28*

Tens and ones

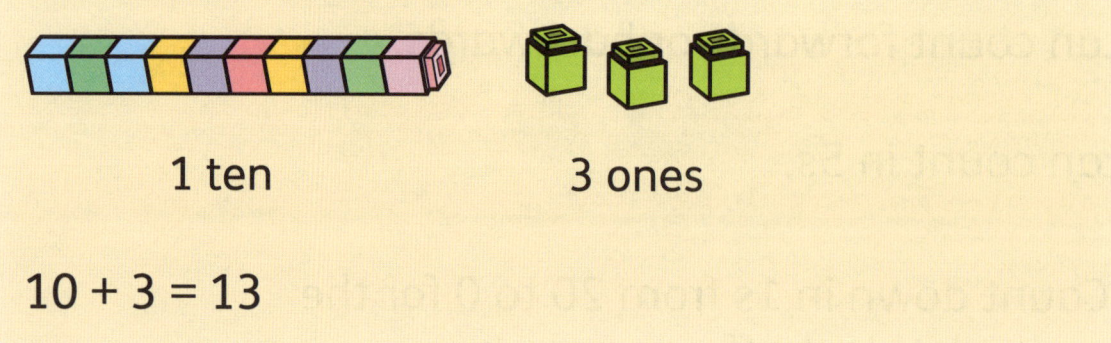

1 ten   3 ones

10 + 3 = 13

1 Say how many tens and how many ones.
Then say the number.

a

b

c

d

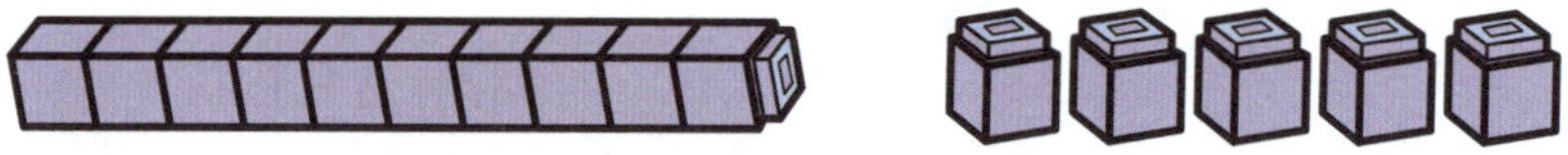

e

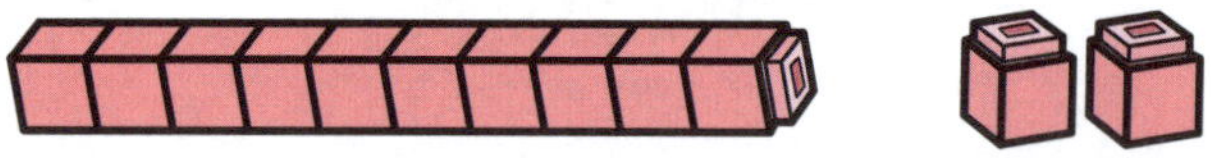

f

g

h

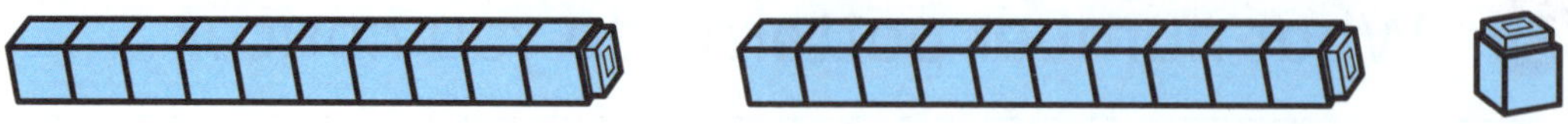

➡ *Workbook page 29 and page 30*

UNIT 7 Mass

Heavier

💭 **Think and share**

The melon is **heavier** than the orange.

The orange is **lighter** than the melon.

Which fruit is more difficult to carry?

Are big things always heavier than small things?

1 Which is heavier?

a   or

b   or

c   or

d   or

e   or

f   or

➡ *Workbook page 31 and page 32*

Lighter

We use heavier and lighter to compare things.

The pen is lighter than the book, but heavier than the sharpener.

1 Which is lighter?

a — or —

b — or —

c — or —

d — or —

e — or —

f — or —

2 Look around your classroom.

   a Find three things that are heavier than a pineapple.

   b Find three things that are lighter than a book.

➡ *Workbook page 33*

Talk about mass

We use different words to talk about **mass**.

My black bag is heavier than your red backpack.

My backpack is the **lightest**. It has the **smallest** mass.

This is the **heaviest**! It has the **greatest** mass.

1 Here are some things we find in a kitchen. Use the words to describe their mass.

mass heavier lighter smaller **greater**

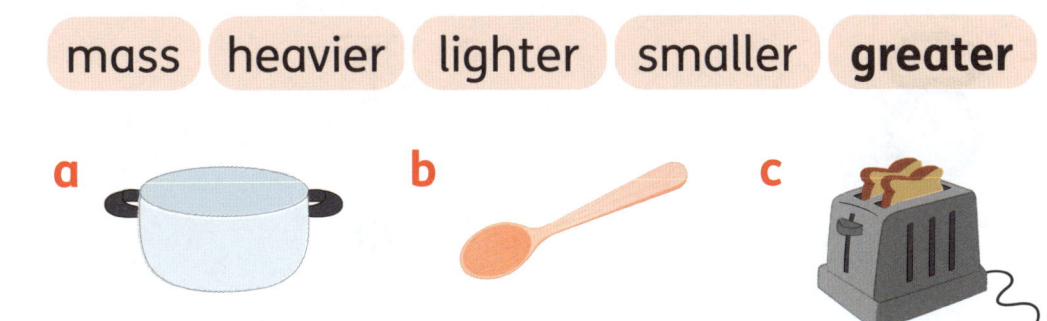

a   b   c

2 Which is heaviest and which is lightest in each group?

a   b

Weighing

A **scale** is an instrument for measuring mass.

Measuring mass is called **weighing**.

1 Where would you find each kind of scale? Discuss with a partner. Use the words to help you.

kitchen   supermarket   bathroom

doctor's office   airport   grocery store

a

b

c

d

e
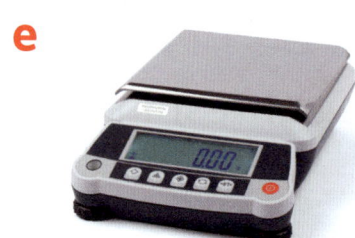

f

How much does it weigh?

Your teacher will bring some fruits and vegetables for you to weigh.

You can also try this at home.

1 You will need:

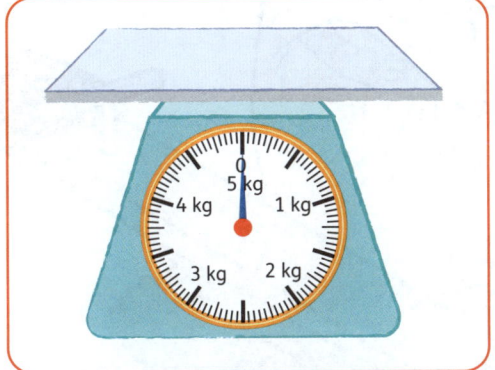

a kitchen scale   fruits or vegetables

- Choose five fruits or vegetables to weigh.
- Write their names.
- Guess which will be heaviest and which will be lightest.
- Weigh.
- Number them from 1 to 5, lightest to heaviest.

➡ *Workbook page 34*

Solve mass problems

Nana

Raj

Mishka Jamila

1 Mishka and Jamila weigh the same.
Raj weighs **more than** Jamila, but **less than** Nana.
What will the seesaw look like with:

a Mishka on one end and Jamila on the other end

b Raj on one end and Nana on the other end

c Mishka and Nana on one end, and Raj and
Jamila on the other end?

Think about who
will be heavier.

Mixed practice 1

1 Say which is **longer**.

a    **b**

2 Say which is full and which is empty.

a    **b**

3 Say something that holds:

a less than a cup    **b** more than a jug

4 Match each statement to the correct measuring instruments.

It measures how long something is.

It measures how much something weighs.

It measures how much something holds.

a    **b**

c    **d**

5 Say these numbers without counting.

a 　　b 　　c 　　d

6 Count.
Make the same number with your cubes
or counters.

a

b

c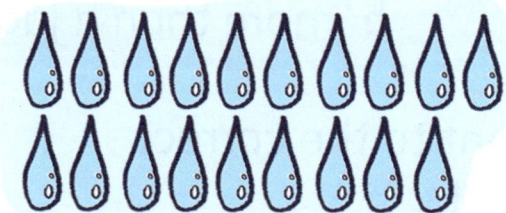

7 Draw a square.
Draw a circle on top and a triangle underneath.

8 Write the numbers from 15 to 20.

9 Say or draw an object that is heavier than
a banana.

10 Say or draw an object that is lighter than
a book.

Add and take away

Add to make 10

Think and share

I'm adding water to the pool.

I'm adding compost to the soil.

I'm adding eggs to the mixture.

Think of other times you add something.

What does add mean in maths?

1 Draw four circles.
Draw two more.
How many circles altogether?

2 Draw three stars.
Draw one more.
How many stars altogether?

lesson continues ⏵

3 Which numbers are added together to make 5 in each picture?
Say the addition.

> We can say an addition like this: One plus four equals five or one add four equals five.

a $1 + 4 = \underline{5}$

b $0 + \underline{} = \underline{}$

c $\underline{} + \underline{} = \underline{}$

4 Which numbers are added together to make 10 in each picture?

> We say an addition like this: One add nine equals ten.

a $1 + 9 = \underline{}$

b $2 + 8 = \underline{}$

c $3 + 7 = \underline{}$

d $\underline{} + \underline{} = \underline{}$

e $\underline{} + \underline{} = \underline{}$

➡ *Workbook page 35*

Count on

Start at 4.

Count on 2 more.

$4 + 2 = 6$

1 Use the number track to count on.

1	2	3	4	5	6	7	8	9	10

a $1 + 2$

b $4 + 3$

c $3 + 1$

d $5 + 4$

e $3 + 2$

f $6 + 3$

g $4 + 1$

h $4 + 6$

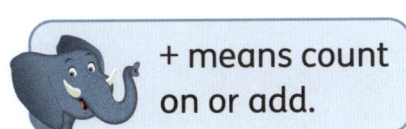

+ means count on or add.

→ Workbook page 36

 Add

3   2   5

1 Count the groups.
Add.
Say the addition.

a

b

c

d

e

➡ *Workbook pages 37 to 41*

Totals

The lightning split each cloud into two numbers.

What is the **total**?

17 ⚡ 3

$17 + 3 = 20$

Total means the number we get when we add numbers together.

1 Work out the total.
You can use counters to help you.

a 6 ⚡ 13

b 11 ⚡ 4

c 8 ⚡ 8

d 7 ⚡ 12

e 15 ⚡ 5

f 7 ⚡ 8

g 10 ⚡ 6

2 Match the clouds that have the same totals.

Count back

Start at 5.

Count back 3.

$5 - 3 = 2$

1 Use the number track to count back.

1	2	3	4	5	6	7	8	9	10

a 6 – 1 ●●●●●○

b 5 – 2 ●●●○○○

c 7 – 3 ●●●●○○○○

d 4 – 2 ●●○○

e 2 – 1 ●○

f 8 – 5 ●●●○○○○○

We can say the subtraction like this: Five take away three equals two.

➡ *Workbook page 42*

Take away

– means **take away** or count back.

3 tomatoes.

Take away 1.

$3 - 1 = 2$

2 left.

You can also use counting back.

1 Take away.

a

$4 - 1 =$ _____

b
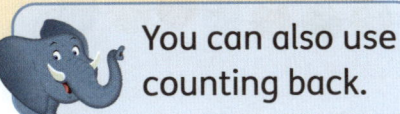
$5 - 2 =$ _____

c

$6 - 1 =$ _____

d

$7 - 5 =$ _____

e

$5 - 4 =$ _____

f

$8 - 2 =$ _____

1	2	3	4	5	6	7	8	9	10

➡ *Workbook page 43 and page 44*

Add and take away

You can add and take away to solve problems.

**Problem solving**

A bracelet has 8 beads.

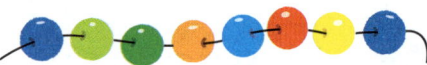

1 How many must you add to make the bracelet?

a $6 + \underline{\quad} = 8$

b $3 + \underline{\quad} = 8$

c $4 + \underline{\quad} = 8$

2 How many must you take away to make the bracelet?

a

$9 - \underline{\quad} = 8$

b

$11 - \underline{\quad} = 8$

c

$14 - \underline{\quad} = 8$

Count back to work out how many to take away.

➡ *Workbook page 45*

Time

Times of day

Think and share

Some things happen during the day.
Some things happen at night.

What happens in the morning?
What happens in the evening?

Talk about things you do every
day or every night.

1 Draw the diagram.
Draw pictures in the correct places.

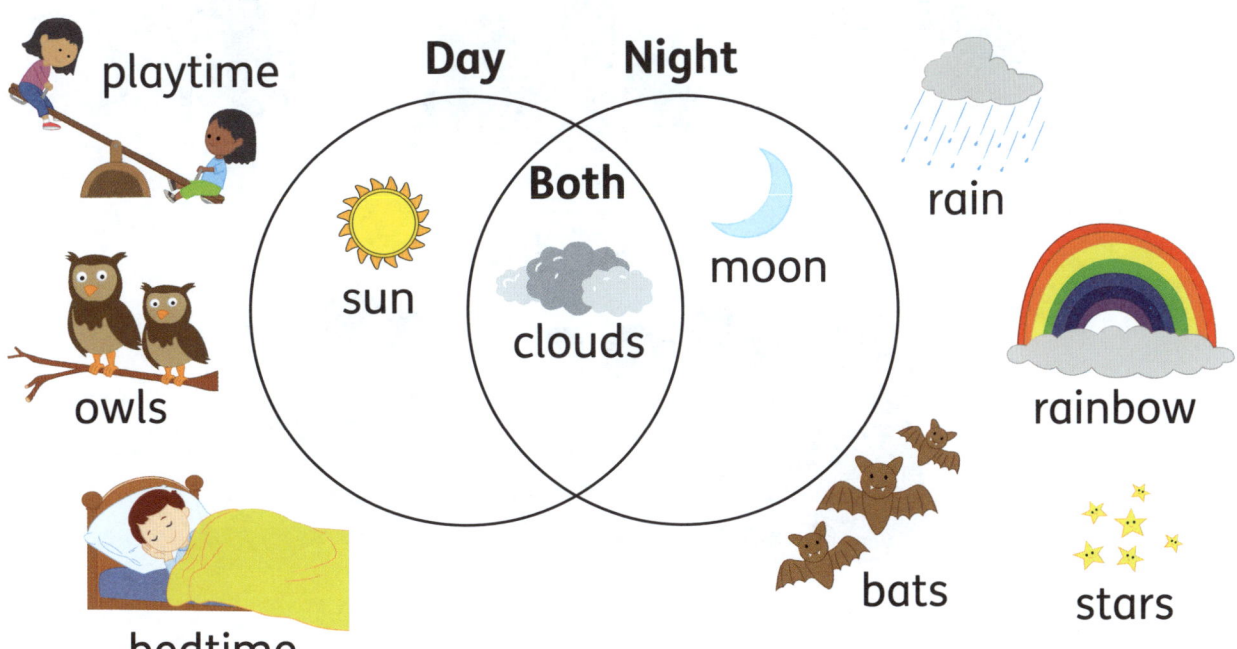

playtime

owls

bedtime

Day **Night**

Both

sun

clouds

moon

rain

rainbow

bats

stars

➡ *Workbook page 46 and page 47*

Days of the week

There are seven **days** in the **week**.

1 Say what Federico did on each day of this week.

Monday

Tuesday

Wednesday

Thursday

Friday

Saturday

Sunday

➡ *Workbook page 48*

Quicker and slower

It is quicker to make toast than to bake a cake.

It is slower to bake a cake than to make toast.

1 Say which is quicker and which is slower.

a    or

tying a shoelace   building a tower

b    or

eating a sandwich   eating a grape

c    or

running across   walking across
the field   the field

2 What do you do that is quick?
What do you do that is slow?

Talking about time

A month is about 4 weeks, or about 30 days.
A **year** is 12 months.

The **months** of the year are January, February, March, April, May, June, July, August, September, October, November, December. A calendar shows all the **dates** in a month.

Look at the calendar. Answer the questions.

MAY						
Sunday	Monday	Tuesday	Wednesday	Thursday	Friday	Saturday
	1 ☺ first day of the month	2	3	4	5	6
7	8	9	10	11	12	13 beach trip
14	15	16 funfair	17	18	19	20
21	22	23	24	25 Mum's birthday	26	27
28	29	30	31			

lesson continues ●

1 Which day of the week is it?

a 😊 the first day of the month

b 🎡 funfair

c 🎁 Mum's birthday

d ⛱️☀️ beach trip

2 What is the day **after** the first day of the month?

3 What day of the week is:

a the last day of May

b 17th of May

c 29th of May?

4 How many Mondays are there in May?

5 Which dates in May fall on a Thursday?

6 What is the month before May?

7 What is the date one week after Mum's birthday?

8 Look at the calendar for May of this year with your teacher. What is the same? What is different?

💡 **Problem solving**

9 I am thinking of a day on this calendar. The date has tens but no ones. It is in the last week of the month. What is the day?

Which dates only have tens?

➡ *Workbook page 49*

Hours, minutes and seconds

When a clock ticks, each tick is one **second**.

There are 60 seconds in one **minute**.

60 minutes make 1 **hour**.

1 Listen to the seconds on a ticking watch or clock. Count 20 seconds.

2 How many can you do in one minute?

skip

jump

draw shapes

3 Which activity takes about an hour?

having a shower

making a cake

➡ *Workbook page 50*

UNIT 10 More adding and taking away

Make 10

Think and share

A group of 9 crayons and 1 more crayon.

$9 + 1 = 10$

How many different ways can you make 10?

1 A full box has 10 crayons. How many more crayons does each box need? Copy and complete.

a $5 + \_\_ = 10$ **b** $8 + \_\_ = 10$ **c** $4 + \_\_ = 10$

2 How many crayons have been taken away from each full box? Copy and complete.

a $10 - \_\_ = 7$ **b** $10 - \_\_ = 6$ **c** $10 - \_\_ = 4$

➡ *Workbook page 51, page 52 and page 53*

Fact families

When we add two numbers together to make a larger number, we can make a family of + and – facts.

$2 + 3 = 5$ $5 - 3 = 2$ $3 + 2 = 5$ $5 - 2 = 3$

1 Make a family of facts for each picture.

a

b

c

d

e

➡ *Workbook page 54 and page 55*

Make 20

16 + 4 = ?

	10	1 ten
	6 + 4 = 10	1 ten
	2 tens = 20	

10 + 6 = 16   16 + 4 = 20

1 Use tens to help you work out the totals.

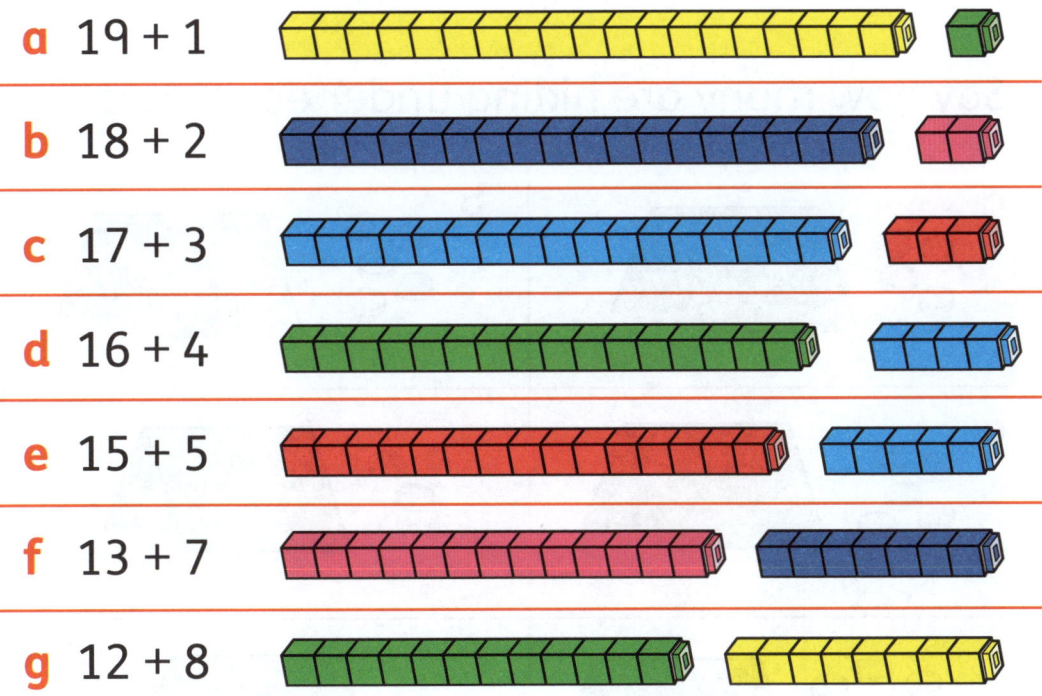

a 19 + 1

b 18 + 2

c 17 + 3

d 16 + 4

e 15 + 5

f 13 + 7

g 12 + 8

2 What pattern do you see in the additions to 20?

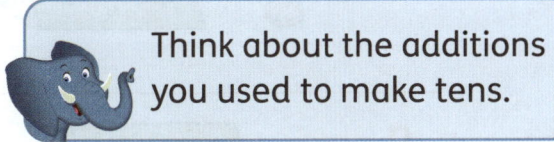

Think about the additions you used to make tens.

3 Choose two additions from question 1. Use them to make a fact family with two + facts and two – facts.

What's missing?

There are 20 counters in each set.

Some are hiding under the card.

20 − 8 = 12

1 Say how many are hiding under each card.

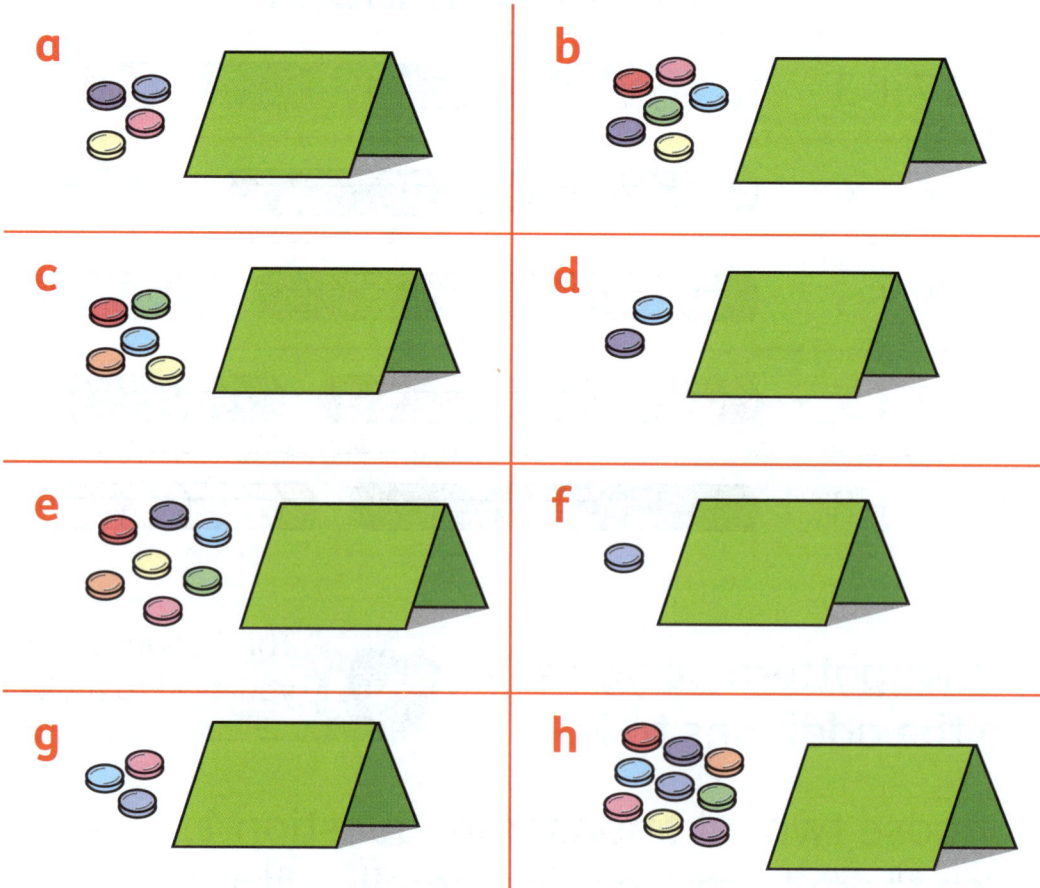

a

b

c

d

e

f

g

h

Find the difference

What is the **difference?**

The chocolate ice cream has 1 more scoop.

$3 - 2 = 1$

The difference is 1.

3 scoops   2 scoops

1 What is the difference?

a

The taller stack has ☐ more.

$7 - 5 = ☐$

b

The smaller pile has ☐ fewer books.

$6 - 3 = ☐$

 **Problem solving**

2 Nico is 6 years old and his sister is 15.
What is the difference in their ages?

 To find the difference you can take the smaller number away from the bigger number.

Full, half full or empty

 **Think and share**

Capacity means how much something holds.

How do you know when a container is full?

How do you know when it is empty?

1 Point to each object. Say if each one is full, half full or empty.

a

b

Which holds more?

Sometimes you can see which container can hold more.

1 Which one can hold more?

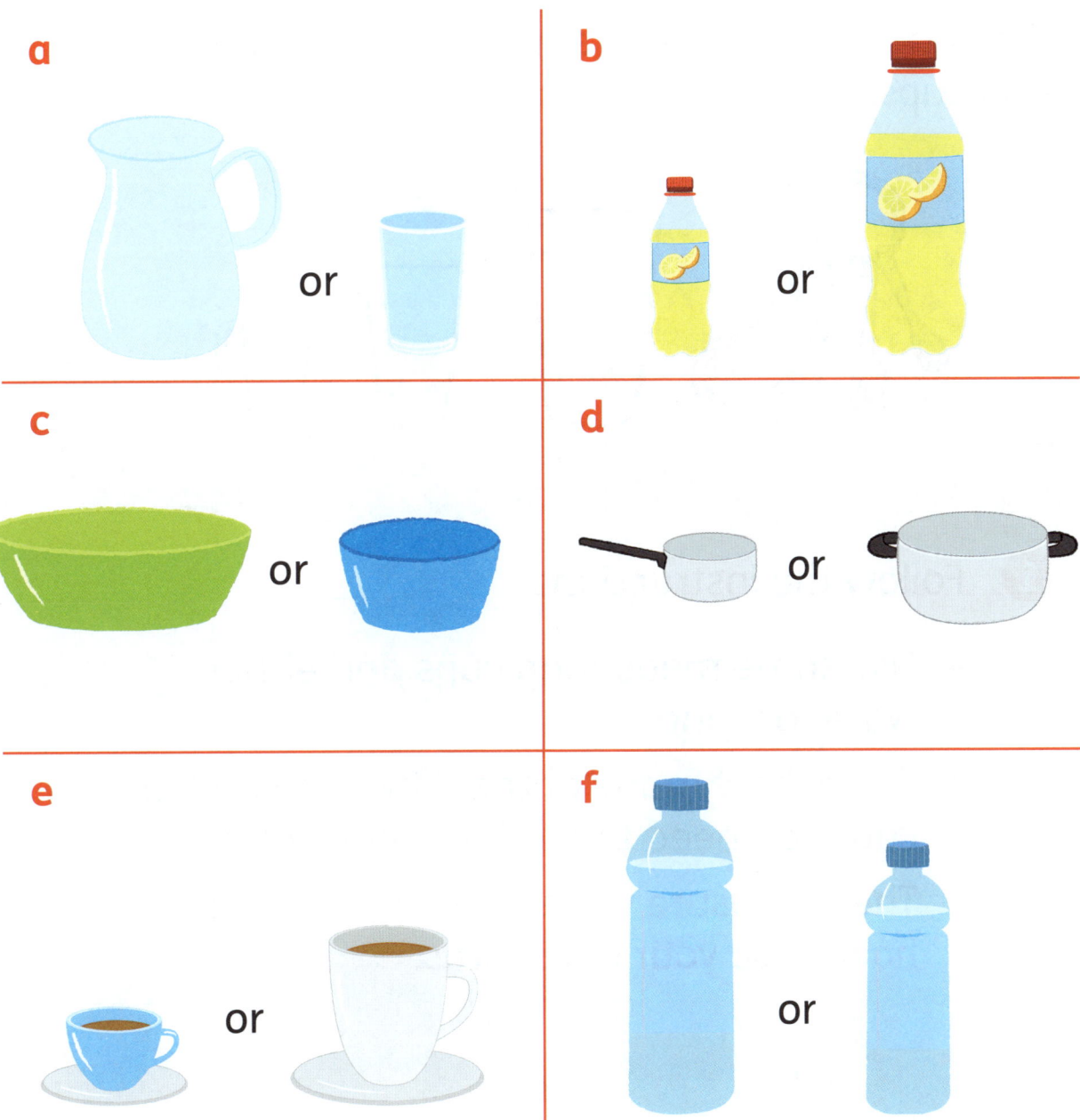

a

or

b

or

c

or

d

or

e

or

f

or

➡ *Workbook page 56*

Estimate and measure

How much does it hold?

We can **estimate**. We make a guess based on what we know.

We can measure. We use measuring spoons and cups.

1 Follow the instructions.

- Use some measuring cups and either water or sand.

- Guess how many of the different smaller cups you need to fill the biggest one.

- Then measure water or sand and see how close your guess was.

➡ Workbook page 57 and page 58

Hot or cold

cold weather
warm clothes

hot weather
cool clothes

1 Warm clothes or cool clothes?
Cold weather or hot weather?

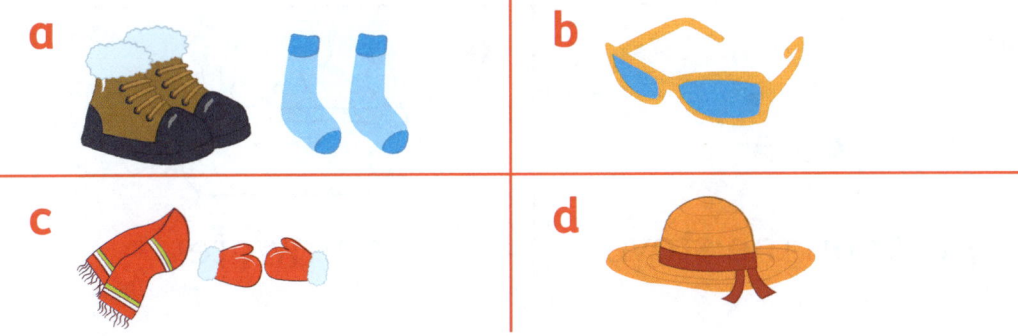

a

b

c

d

2 Are these things hot or cold?

a

b

c

d

e

f

How warm or cool is it?

We use different words to say how hot or cold things are.

icy cold   cold   cool   warm   hot   very hot

1 How does the water feel?

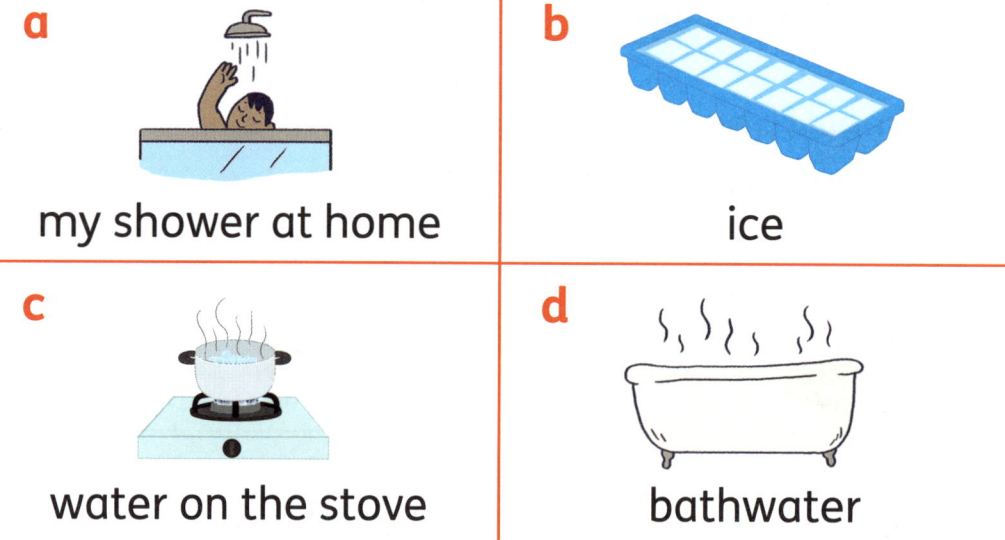

a   my shower at home

b   ice

c   water on the stove

d   bathwater

2 How does the air feel?

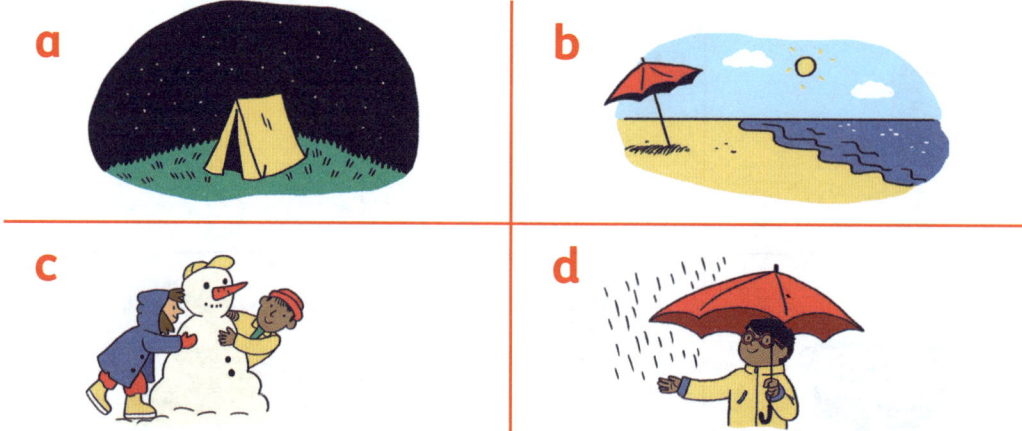

a

b

c

d

3D shapes

Name 3D shapes

3D shapes take up space.

| sphere | cylinder | cone | cube | cuboid | pyramid |

Balls, boxes and tins are all 3D shapes.

1 Which 3D shapes can you find in the picture?

2 Look at the shapes in the picture.
Which shapes can you see that:

a stack   **b** roll   **c** slide?

➡ *Workbook page 59 and page 60*

Describe 3D shapes

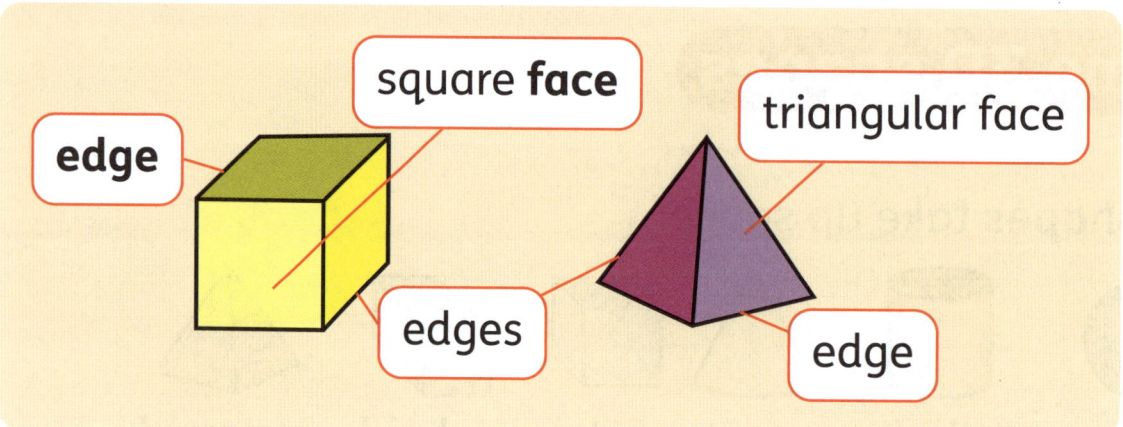

1 Look at the pictures and answer the questions.

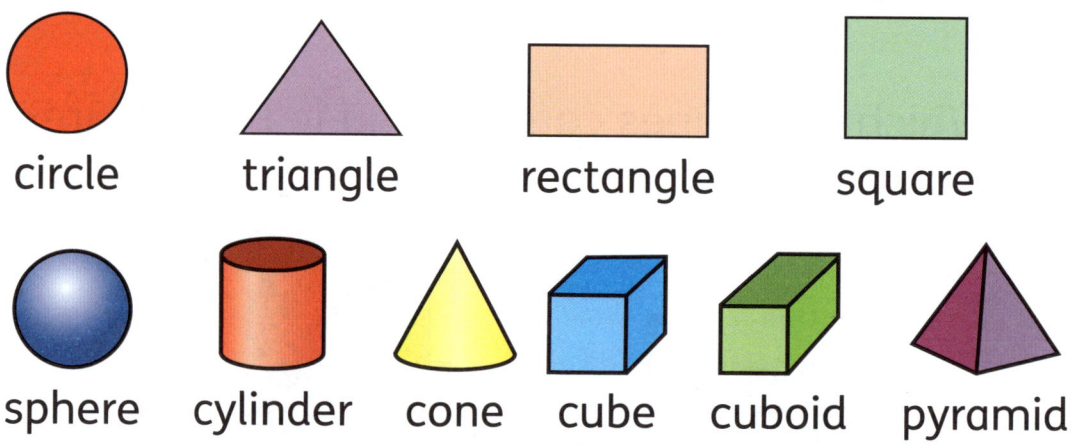

circle   triangle   rectangle   square

sphere   cylinder   cone   cube   cuboid   pyramid

a What makes 3D shapes different from 2D shapes?

b Which 3D shapes have circles as faces?

c Which 3D shape has no edges?

d Which 3D shape has only square faces?

e Which 3D shape has triangular faces?

f Which 3D shapes have six faces?

Remember, circles, triangles, squares and rectangles are 2D shapes.

➡ *Workbook page 61*

Alike and different

Look at the cube and the cuboid.

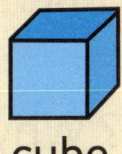
cube

cuboid

They are alike in some ways.

- They both have 6 faces and 12 edges.
- Each face has 4 straight sides and 4 square corners.

They are different in some ways.

- The cube only has square faces, but the cuboid has rectangular faces.

1 Say what is alike and what is different about each pair of 3D shapes.

a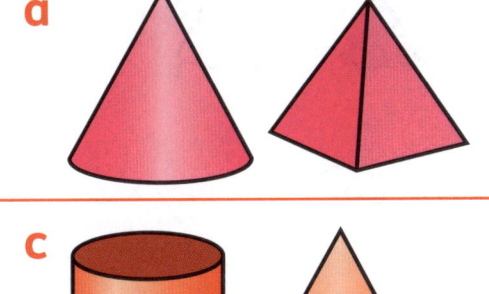

b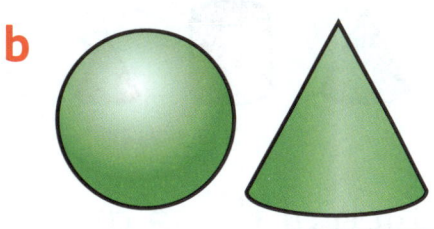

c

d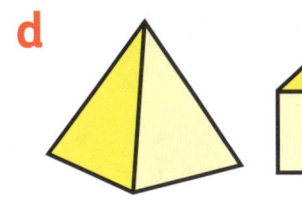

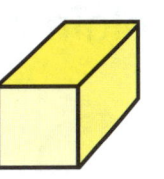

Which 3D shape doesn't match?

Find one 3D shape in each group where the shape, faces or edges are different from the others.

1 Which 3D shape doesn't match? Why?
Talk about the names of the shapes, the number of faces and the number of edges.

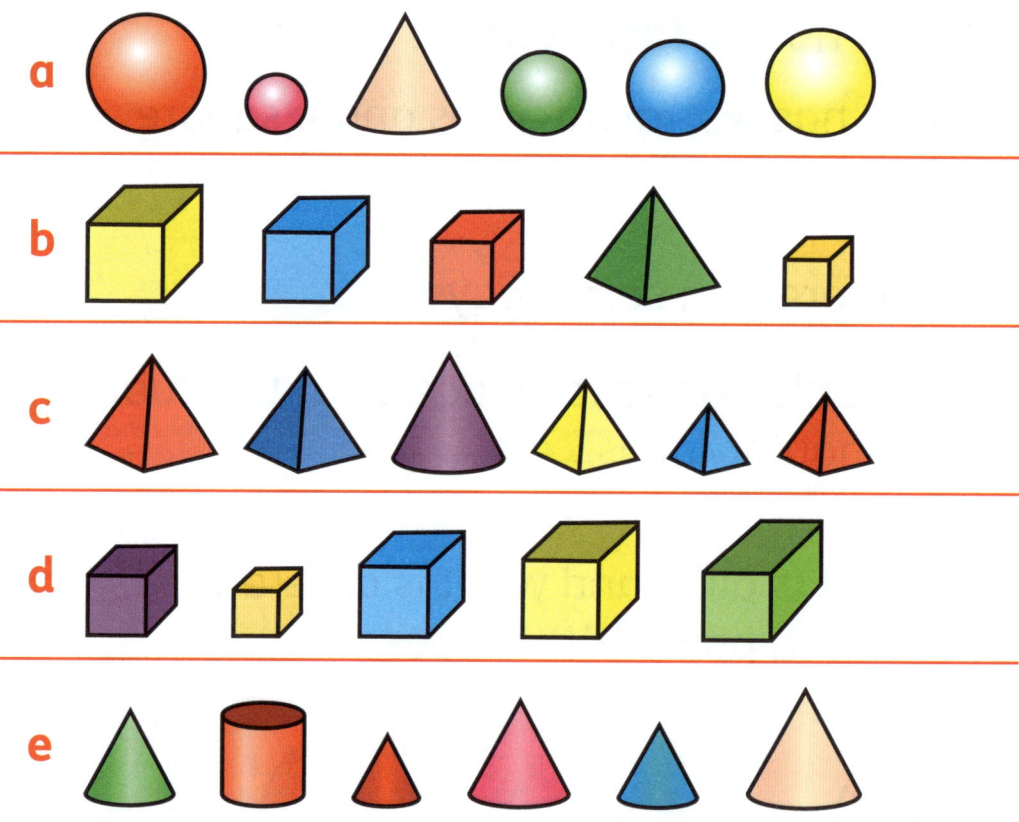

2 Which of these are not spheres?
Explain why not.

a    b    c    d   e    f

Make rows and columns

This is an **array**.

2 **rows**. 3 frogs in each row.

3 **columns**. 2 frogs in each column.

- 3 + 3 = 6   add the rows

- 2 + 2 + 2 = 6   add the columns

You can put objects on squared paper like this:

You can shade squared paper like this:

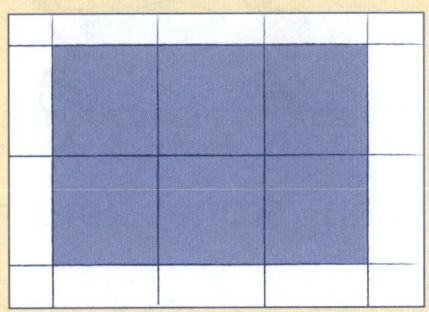

1 Use squared paper. Make arrays to show these numbers.

a 4   **b** 8   **c** 9

➡ *Workbook page 62*

Skip count

Mawusi jumps 3 stones with each skip.

She skips 4 times.

1 skip	3 stones
2 skips	6 stones
3 skips	9 stones
4 skips	12 stones

1 Ben jumps 4 stones with each skip.
He skips 4 times.

How many stones does he jump in total?

2 Alia jumps 5 stones with each skip.
She skips 3 times.

How many stones does she jump in total?

Multiply

When you **multiply** a number, you add it a number of times.

This is called **repeat adding**.

Each card has 2 dots.

2 times: ●● + ●● = 4 dots

3 times: ●● + ●● + ●● = 6 dots

4 times: ●● + ●● + ●● + ●● = 8 dots

1 How many dots altogether?

a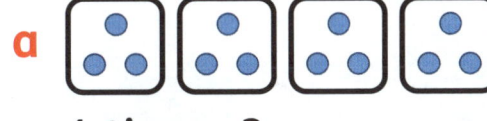

4 times 3

b

3 times 5

c

6 times 1

d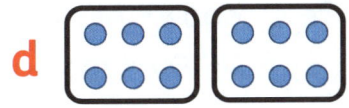

2 times 6

e

3 times 4

f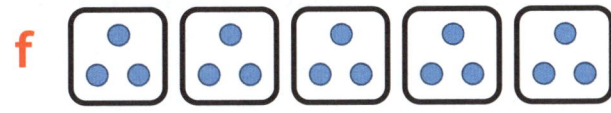

5 times 3

Doubles

If you pick a domino, you can sometimes pick a **double**, like this.

Amy picked a double 5.   Layla picked a double 2.

5 + 5 = 10   2 + 2 = 4

double 5 = 10   double 2 = 4

What does it mean when you double a number?

1 Say and write the number double each domino shows.
Then write the total number of dots each domino has.

a    b    c

d    e    f

💡 **Problem solving**

What makes a number odd and what makes a number even?

2 5, 7 and 9 are all odd numbers.
When you double an odd number, is the total odd or even? Why?

➡ *Workbook page 63*

Arrays

Look at this array of dots.

2 rows

4 dots in each row

$4 \times 2 = 8$

four times two
equals eight

or

$2 \times 4 = 8$

two times four
equals eight

1 Say how many rows.
Say how many dots are in each row.

Use the words times and equals to
describe each array.

a

b

c

d

e

f

 **Problem solving**

 Look at the number of dots
and the number of rows.

2 When do the dots make a square?

Different ways to multiply

You have learnt different ways to multiply.

$4 + 4 + 4 = 12$

repeat add

use an array

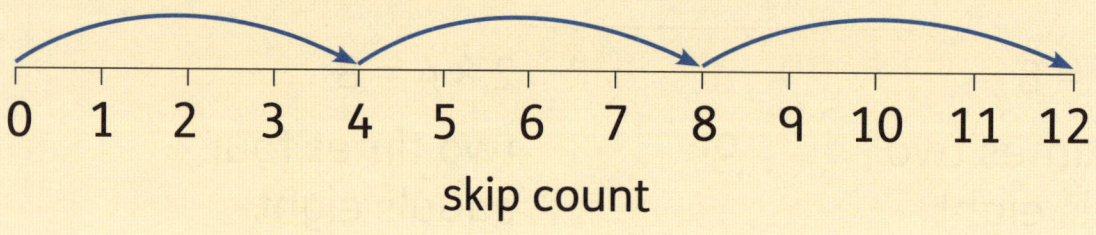

skip count

1 Multiply. Copy and complete the calculation.

a

2 jars

4 marbles in each

$2 × 4 = \underline{\quad}$

b

4 cakes

5 candles on each

$4 × 5 = \underline{\quad}$

c

2 plates

6 cherries on each

$2 × 6 = \underline{\quad}$

d

3 bowls

5 balls in each

$3 × 5 = \underline{\quad}$

➡ *Workbook page 64*

Zero to 100

Larger numbers

Think and share

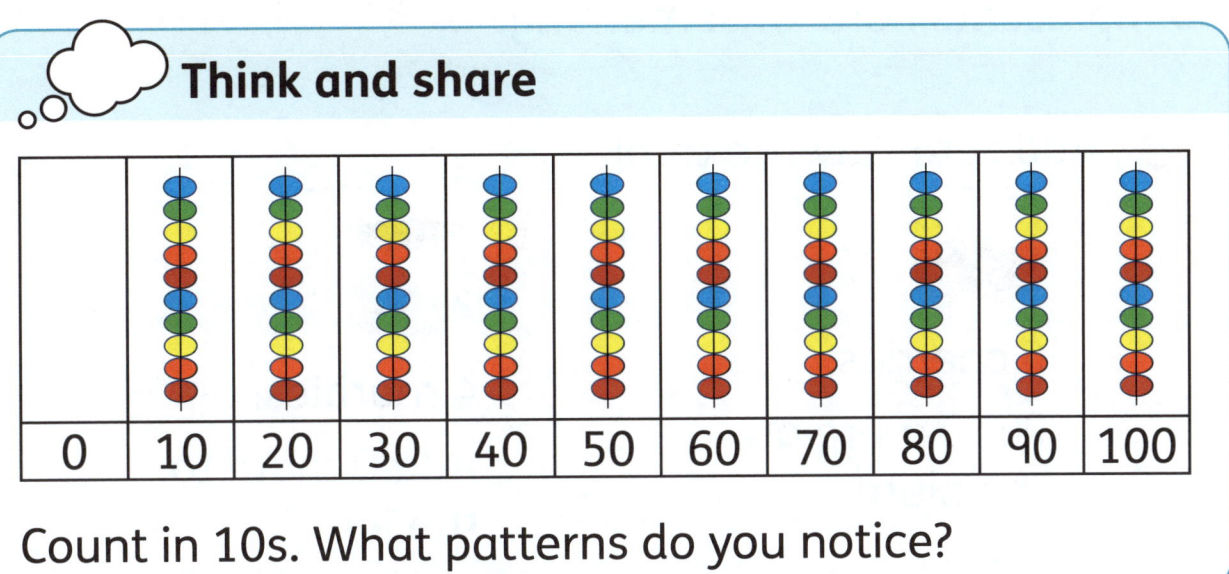

| 0 | 10 | 20 | 30 | 40 | 50 | 60 | 70 | 80 | 90 | 100 |

Count in 10s. What patterns do you notice?

1 Match the number names to numbers on the **chart**.

ninety   twenty   eighty   forty

sixty   fifty   zero   thirty

ten   seventy   one hundred

Problem solving

2 Raj's little sister counts like this:

Twenty, thirty, forty, fivety, sixteen, seventeen, eighteen, ninety, twenty!

Explain her mistakes.

➡ *Workbook page 65*

 **Zero**

0 is **zero**. Another word for zero is **nought**.

0 represents none of something.

1 Add zero. Say the total.

a

6 cherries.

Put 0 more on the plate.

6 + 0 = ___

b

4 marbles.

Add 0 more to the jar.

4 + 0 = ___

2 Take away zero. Say how many are left.

a

8 pencils.

Take away 0.

8 – 0 = ___

b

5 candles.

Take 0 off the cake.

5 – 0 = ___

 **Problem solving**

3 I have a plate of sandwiches. All the sandwiches get eaten. How many are left?

➡ *Workbook page 66*

Number lines and charts

We can use number lines and number charts to help us count to larger numbers.

1 Use the number lines to count.

a Count in 1s from 20 to 30.

First count forwards.

Then count backwards.

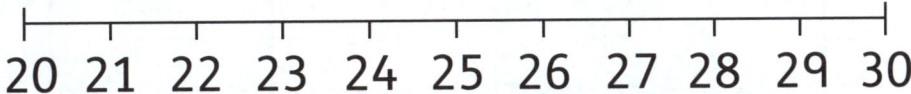

20 21 22 23 24 25 26 27 28 29 30

b Now count from 30 to 40.

First count forwards.

Then count backwards.

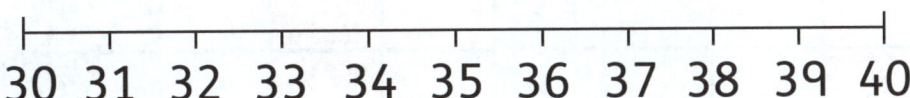

30 31 32 33 34 35 36 37 38 39 40

c What was the same?

What was different?

2 Which numbers are missing from this number line?

26   28 29   31 32 33   35

lesson continues ⊙

3 Work with a partner and take turns.
Pick a number.
Count ten numbers forwards.
Then count ten numbers backwards.

4 Here is a number chart to 100.
What patterns do you notice?

1	2	3	4	5	6	7	8	9	10
11	12	13	14	15	16	17	18	19	20
21	22	23	24	25	26	27	28	29	30
31	32	33	34	35	36	37	38	39	40
41	42	43	44	45	46	47	48	49	50
51	52	53	54	55	56	57	58	59	60
61	62	63	64	65	66	67	68	69	70
71	72	73	74	75	76	77	78	79	80
81	82	83	84	85	86	87	88	89	90
91	92	93	94	95	96	97	98	99	100

➡ *Workbook page 67 and page 68*

Count on a number chart

1	2	3	4	5	6	7	8	▲	10
11	12	13	14	15	16	17	18	19	20
21	22	⬠	24	25	★	27	28	29	30
31	32	33	34	35	36	◇	38	39	■
✹	42	43	44	45	46	47	48	49	50
51	52	53	54	55	56	57	58	59	60
61	62	63	✸	65	66	67	⬡	69	70
71	72	73	74	75	76	77	78	79	80
81	●	83	84	85	86	87	88	89	90
91	92	93	94	95	96	97	⬭	99	100

1 Some numbers on the 100 chart are covered.

   a Start at ■ and count forwards up to 45.

   b Start at ▲ and count backwards to 1.

   c Start at ● and count forwards to 100.

2 Say which other numbers are covered
by shapes.

3 Work with a partner.
Give your partner instructions. Tell them:

- which number to start on
- whether to count forwards or backwards
- which number to finish on.

Rocket countdown

1 On each rocket, count backwards from the greatest number to the smallest number. Say the missing numbers.

a

28
27
26

23

21
20

b

44
43

41

39

37

c

60

58
57

55

d

100
99

96

 **Problem solving**

2 I am a number with two digits.

- My second digit is bigger than my first digit.
- If you add my digits, they make 8.
- The difference between my digits is 4.

What number am I?

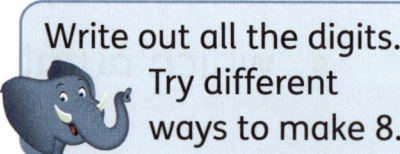
Write out all the digits. Try different ways to make 8.

Fast track, slow track

To play this game you need one small object or counter for each player, and a 1–6 spinner.

- Take turns to spin the spinner.
- If you get an even number (2, 4 or 6), move 10 places.
- If you get an odd number (1, 3 or 5), move 1 place.

1 Play the game in pairs or groups.
The first person to get to 100 wins.

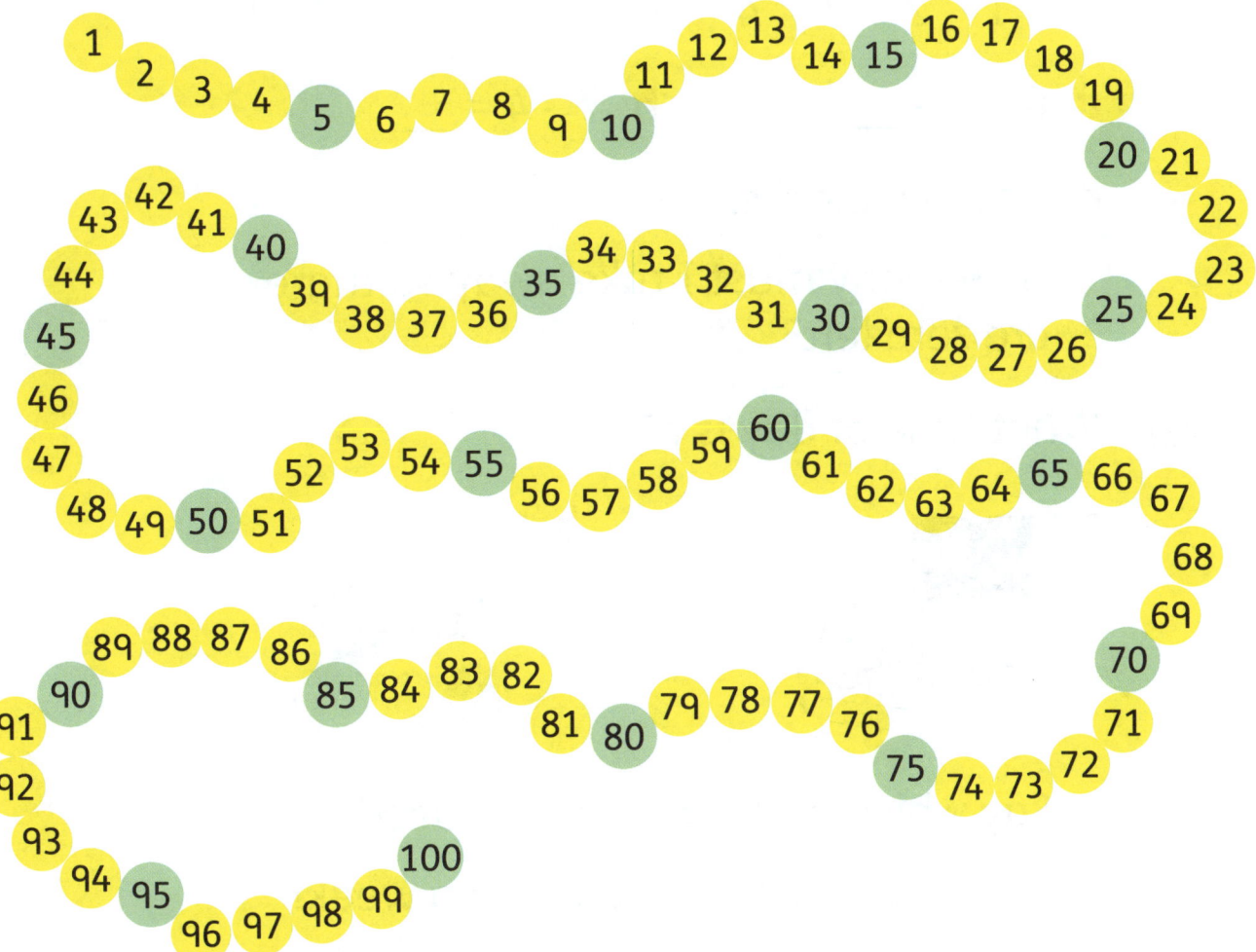

Mixed practice 2

1 Copy and complete these calculations.

a 5 + 2 = ___

🚗🚗🚗🚗🚗 + 🚙🚙

b 4 + 3 = ___

🚗🚗🚗🚗 + 🚙🚙🚙

2 Use the number line to help you take away. Copy and complete these calculations.

```
0   1   2   3   4   5   6   7   8   9   10
```

a 8 – 3 = ___   **b** 7 – 4 = ___

c I have 5 blocks.

How many must I take away so there are zero left?

3 What time of day is it?

a    **b**

4 **a** Say which day comes after Tuesday.

b How many days are there in a week?

5 Add.

    a 1 ten + 4 ones   **b** 1 ten + 5 ones

6 What is the difference between the numbers in each pair?

    a 3 and 5    **b** 10 and 5

Remember, to find the difference, take the smaller number away from the bigger number.

7 Say which container can hold more.

    a  or    **b**  or

8 Name these 3D shapes.

    a   **b**   **c**   **d**

9 Multiply.

    a 4×3    **b** 3×2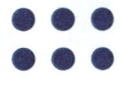

10 Say the missing numbers.

    a 23 ___ 25 26 ___ 28 29 ___

    b 49 48 47 46 ___ ___ 43 42 ___ ___

UNIT 15 Position and direction

Position words

> ### 💭 Think and share
>
> Position words tell us where things are.
>
> Use the position words to say what the picture shows.
>
>
>
> on   in   under   in front of   behind   next to

1 Where is the bird?
Choose the correct position word to fill each gap.

a _____ the tree   **b** _____ the ladder

c _____ the nest   **d** _____ the watering can

e _____ the spade   **f** _____ the bucket

➡ Workbook page 69

Directions

Sometimes we give directions.

Directions tell someone where to go.

Go round the corner.

1 Help the mouse to get the cheese.
Use the direction words.

over	under	into	out of
through	around	up	down

a

b

c

d

e

f

➡ *Workbook page 70*

Forwards or backwards

Sometimes we go forwards.

Sometimes we go backwards.

1 Where must the person or object go? Forwards or backwards?

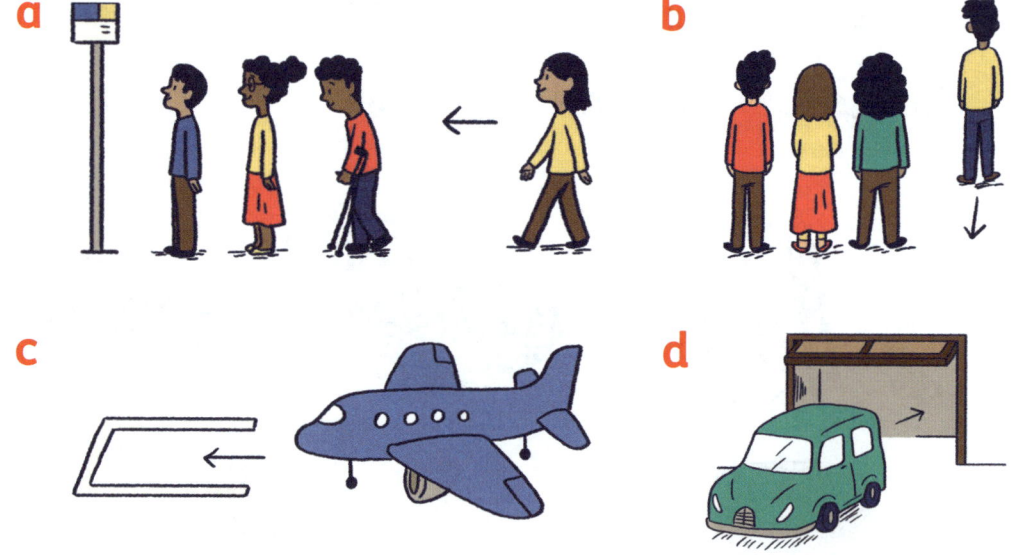

a

b

c

d

2 Choose an object in the room and write down its name. Do not show your partner. Give them directions to get to the object without telling them what it is.

Half turns and full turns

Two **half turns** make a **full turn**.

When you make a full turn, you end pointing the same way that you started.

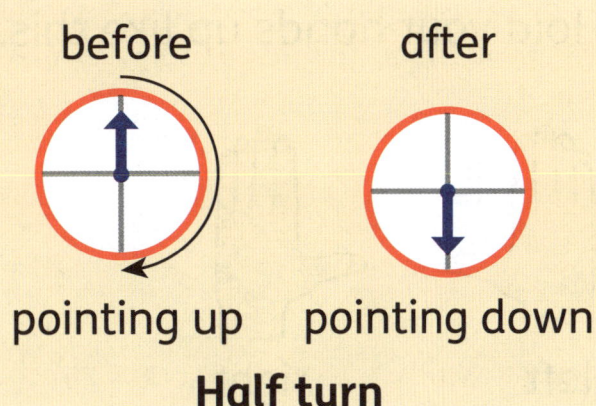

before   after

pointing up   pointing down

Half turn

1 Full turn or half turn?

a
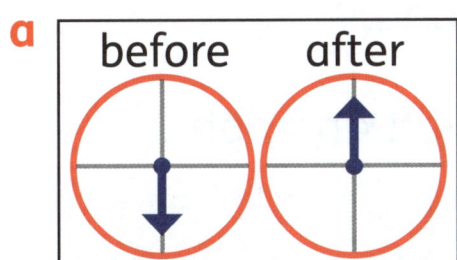

b
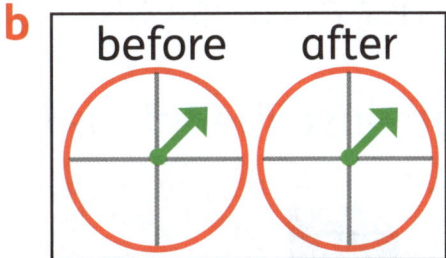

c
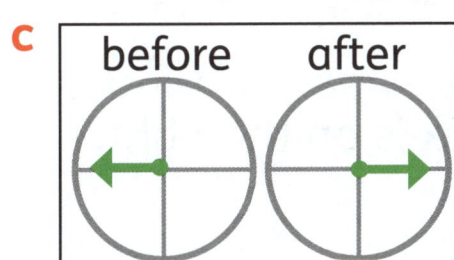

d
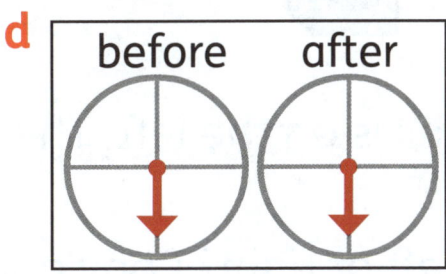

 **Problem solving**

2 I face forwards and make a half turn.

Then I make three full turns.

Which direction am I facing?

Try following the directions yourself.

➡ *Workbook page 71*

Left and right

Hold your hands up like this.

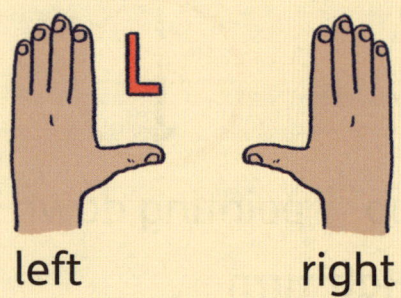

left   right

The hand that makes a L shape is your left hand.

The other hand is your right hand.

We can describe things using left and right.

The ball is on the left. The block is on the right.

1 Left or right? Copy and complete.

The spoon is on the _____.

The pot is on the _____.

lesson continues ◗

2 Left or right? Copy and complete.

a

The pineapple is on the _____.

The melon is on the _____.

b

The bat is on the _____.

The tennis ball is on the _____.

c

The cups are on the _____.

The books are on the _____.

3 Face forwards. What can you see?

Look left. What can you see?

Look right. What can you see?

Part of a turn

Four **quarter turns** make a full turn.

a quarter turn   a half turn   a three-quarter turn   a full turn

We can describe the **size** of the turn – a quarter, half, three-quarter or full turn.

We can also describe the direction of the turn: clockwise ⟳ or anti-clockwise ⟲

1 Does the arrow make a quarter turn, a half turn, a three-quarter turn or a full turn?
Clockwise or anti-clockwise?
Copy and complete each sentence.
One is done for you.

a <u>a quarter</u> turn anti-clockwise

start   finish

b _____ turn clockwise

 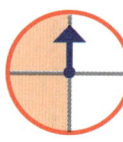

start   finish

lesson continues ⊙

2 Does the arrow make a quarter turn, a half turn, a three-quarter turn or a full turn? Clockwise or anti-clockwise? Copy and complete each sentence.

a _____ turn
anti-clockwise

start   finish

b a quarter turn
_____

start   finish

c _____
clockwise

start   finish

d a quarter turn
_____

start   finish

e _____
_____

start   finish

f _____
_____

start   finish

Problem solving

Use the pictures to help you.

3 Which turns end up in the same position, whether you make them clockwise or anti-clockwise?

UNIT 16 Money

Coins in my country

 **Think and share**

Describe the coins.

Which coins do you use in your country?

Why do we have pictures on coins?

1 Look at some coins from your country.
Trace them into your exercise book.
Say how much each coin is worth.

2 Look at some notes from your country.
Draw them in your exercise book.

Problem solving

3 What can you buy with three coins?

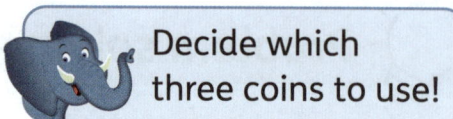 Decide which three coins to use!

➡ *Workbook page 72*

Recognise coins

Here are some coins people use in different countries.

coins from India   coins from Malaysia   coins from Nigeria

Here are some coins people use in the UK.

one penny   five pence   ten pence   twenty pence
1p   5p   10p   20p

fifty pence   one pound   two pounds
50p   £1   £2

1 Look at each coin from the UK.
Say how much it is worth.

a    b    c

d    e    f

2 **a** Which is the smallest coin in your country?

b Which is the biggest?

Make totals

We can add the **value** of a set of coins.

Value is how much it is worth.

 $5p + 5p = 10p$

p means pence.
£ means pound.

1 Count each set of coins. Say how many pounds altogether.

Look at the coin values on page 93 to help you.

2 Work out the total for each set of coins.

Problem solving

3 Find three different ways to make £1 out of smaller coins.

£1 = 100p

➡ *Workbook page 73*

Sort 2D shapes and 3D shapes

Properties of shapes

 **Think and share**

2D shapes and 3D shapes have different **properties**.

2D shapes have lines, sides and corners we can count.

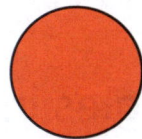  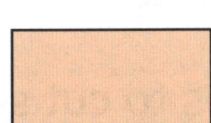 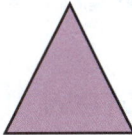

3D shapes take up space. They have faces and edges.

   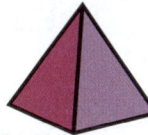  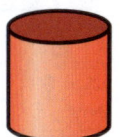

Describe the properties of these shapes.

1 **a** Which shapes are squares and which are rectangles?

A 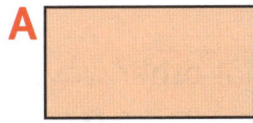 B  C  D  E  F

b What property does a square have that is different from a rectangle?

lesson continues ▶

2 **a** Why are these shapes not rectangles?

 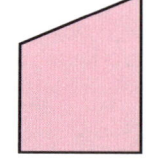 

b Why are these shapes not triangles?

 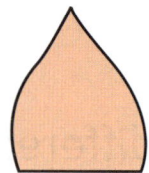 

 **Problem solving**

3 Alex drew lines to cut some rectangles in **half**.

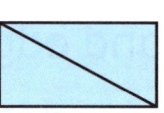 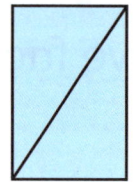 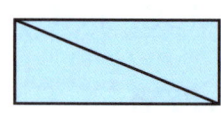

Alex says:

> If you cut a rectangle in half, you always get a triangle.

Do a drawing to show Alex's mistake.

What other ways can we cut a rectangle in half?

➡ *Workbook page 74*

More properties of shapes

We can draw other shapes with curved lines or straight sides.

We only call them circles, triangles, rectangles and squares if they have the correct properties.

1 Say which shapes:
- have only straight sides
- have straight sides and curved lines
- have only curved lines.

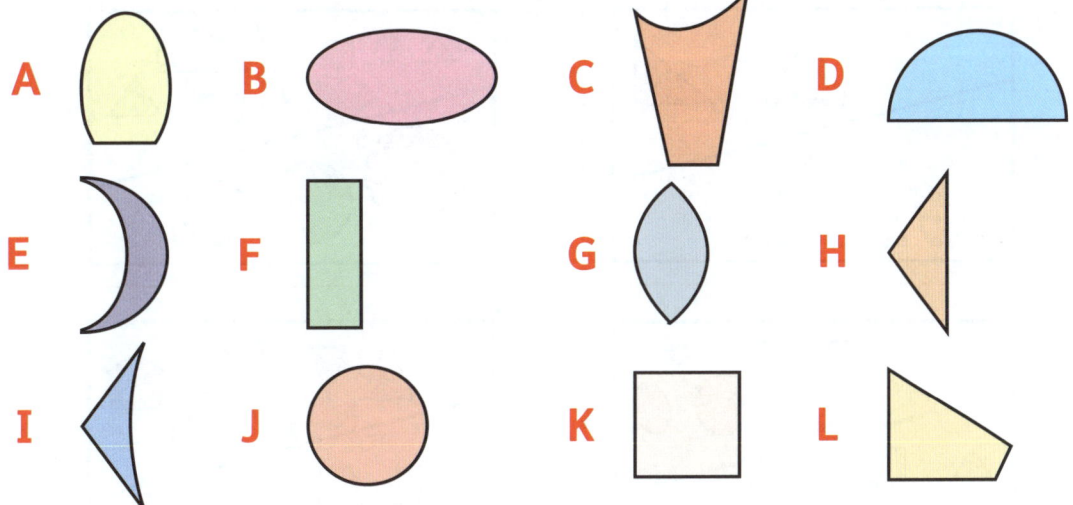

A B C D
E F G H
I J K L

2 The shapes above have straight and curved lines.
Say the number of lines in each shape.

3 Which shapes are rectangles, circles, triangles or squares?
Explain why the others are not any of these.

Sort 2D shapes

We can sort 2D shapes in different ways.

sorted by colour

sorted by number of sides

1 How were these shapes sorted?
There could be more than one answer.

a
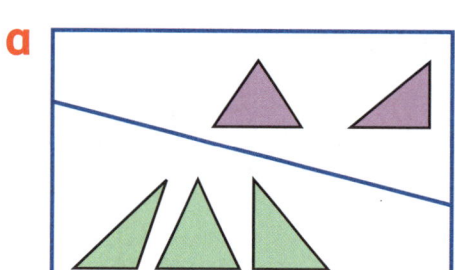

b
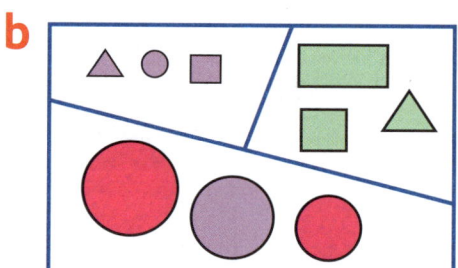

c
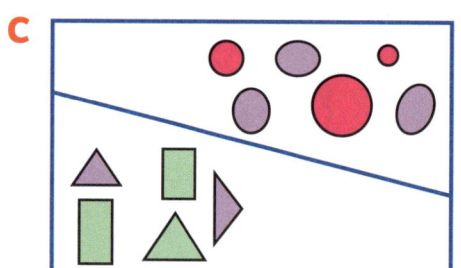

d
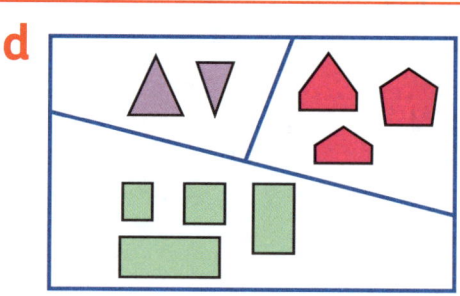

size   straight sides   colour

number of sides

Sort 3D shapes

We can sort 3D shapes by:
- number of faces
- number of edges
- flat faces, curved surfaces, or both
- size
- colour.

1 How were these 3D shapes sorted?

a

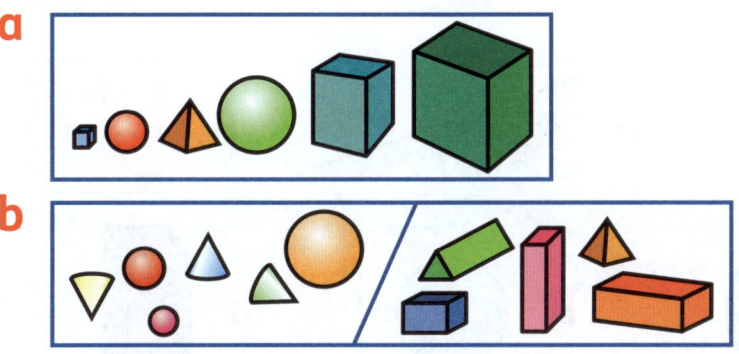

b

by whether it rolls   by size   by colour

 **Problem solving**

2 Name each 3D shape from the description.
  a a 3D shape with no faces
  b a 3D shape that has 1 curved surface
  c a 3D shape with 5 faces
  d a 3D shape with 6 faces

 Use real shapes to help you.

➡ *Workbook page 75 and page 76*

More sorting

1 We can sort these shapes in different ways.

a How many are there of each colour?

b Which are 3D shapes?

c Which have curved lines or surfaces?

d Which only have straight sides or edges?

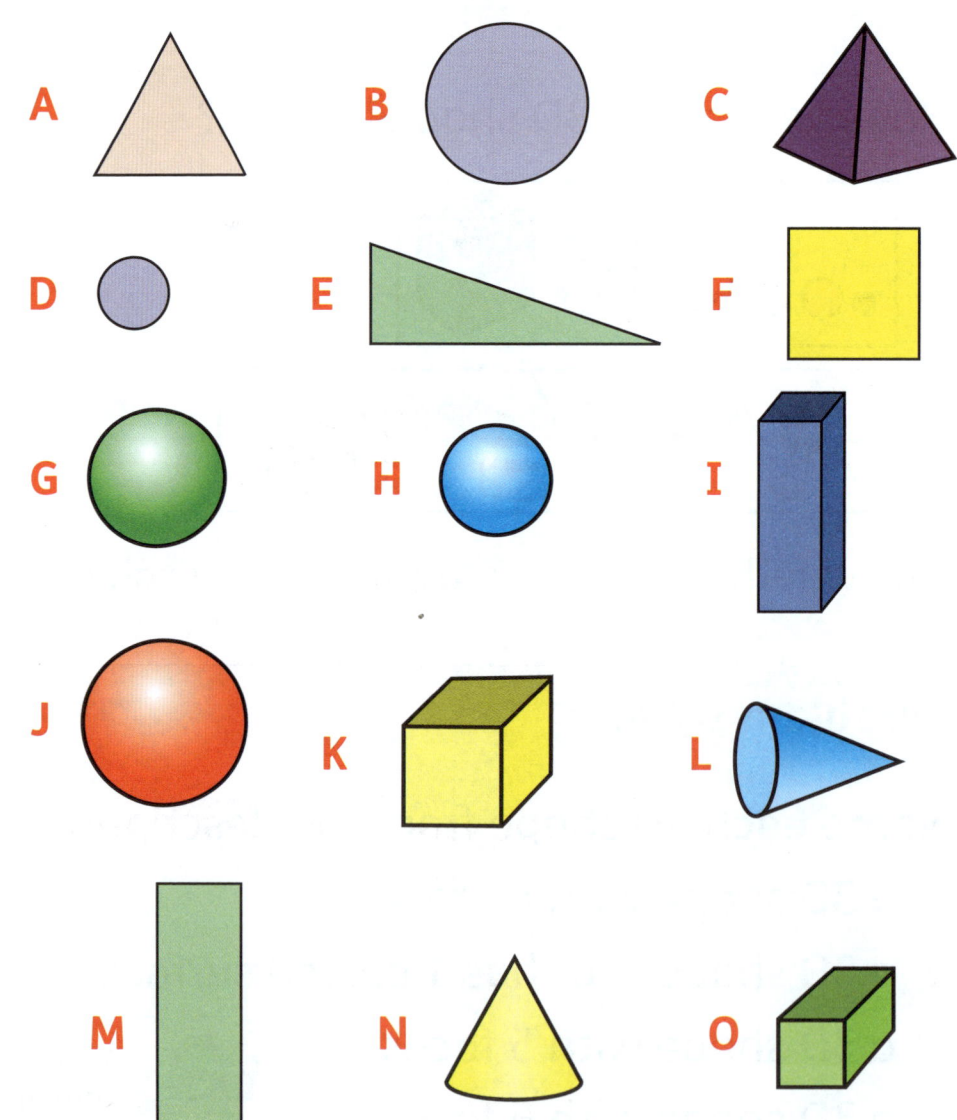

Halves

Think and share

When we cut something in half, both the parts are equal in size. Which cuts show a fair way to share a pizza between two people?

 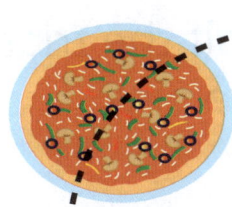 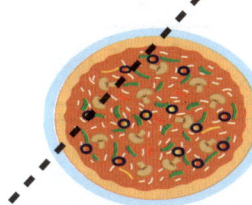 

1 Which are cut in half and which are not?

A 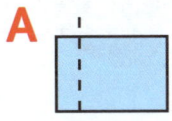   B    C    D

E    F    G    H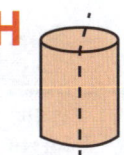

Problem solving

2 Do the shaded parts show halves or not halves? Explain your ideas.

A    B 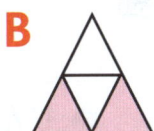   C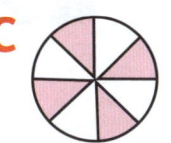

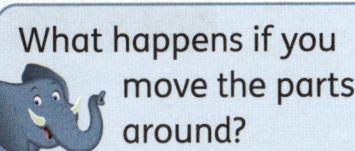

What happens if you move the parts around?

Make halves

A half is one out of two equal parts. We write $\frac{1}{2}$.

Two halves make a **whole**.

We can find half of an object, a shape or a number.

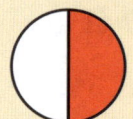 $\frac{1}{2}$ of the circle is shaded.

 Half the triangles are pink.

1 Which are cut into halves?

A   B   C   D

E   F   G   H

2 Half or not half?

a   b   c

→ *Workbook page 77*

Half of a number

Half the 6 candles are lit.
Half of 6 is 3.
We also write half like this: $\frac{1}{2}$.

1 Find half of each number. Copy and complete the calculations.

a

half of 8 = _____

b

half of 12 = _____

c

$\frac{1}{2}$ of 10 = _____

d

$\frac{1}{2}$ of 4 = _____

 **Problem solving**

2 My sister is half my age.
I am 12 years old.
How old is my sister?

Start with my age and then my sister's age.

➤ *Workbook page 78*

Divide

Mia has 6 buns. She wants to share them equally between 2 plates.

When you share something into 2 equal groups, you **divide** it by 2.

The sign for divide is ÷

6 ÷ 2 = 3

Mia can share 6 buns into 2 equal groups of 3.

She wants to give 2 buns to each friend. How many friends can she give to?

When you group something into groups of 2, you divide by 2.

She can give 3 friends 2 buns each.

6 ÷ 2 = 3

1 Mia now wants to give 3 buns to each friend. How many friends can she give to?

lesson continues ●

2 Divide each group into 2 equal groups.
Copy and complete the calculations.

a    b   c

$10 \div 2 =$ _____ $12 \div 2 =$ _____ $8 \div 2 =$ _____

3 How many groups of 3 can you make?
Copy and complete the calculations.

a 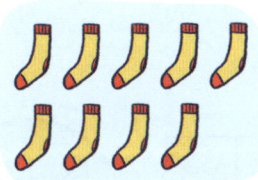   b    c

$9 \div 3 =$ _____ $12 \div 3 =$ _____ $15 \div 3 =$ _____

 **Problem solving**    Use counters or a drawing to help you divide.

4 I have 15 plants. I want to plant them in 3 rows. How many plants will go in each row?

➡ *Workbook page 79*

Quarters

This circle is cut in to four equal parts.

Each part is 1 out of 4 equal parts of the whole circle.

We write $\frac{1}{4}$ or one **quarter**.

Four quarters together make one whole.

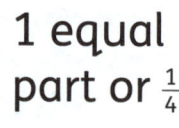
1 equal part or $\frac{1}{4}$

1 Which shapes show quarters?
Point to one quarter.

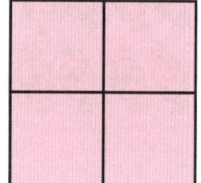 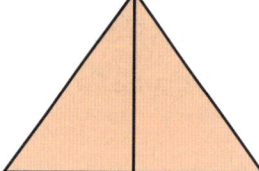 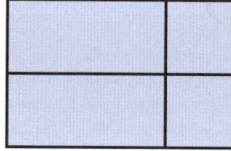 

2 Fold a piece of paper to make quarters.
What are the different ways you can do it?

3 How many quarters make one half?

Quarter of a number

Here are 8 candles.

One quarter of the candles are lit.

$\frac{1}{4}$ of 8 = 2

One quarter is 1 out of 4 equal parts.

1 Work out one quarter of each group.
Copy and complete the calculation.

a $\frac{1}{4}$ of 12 = ___

b $\frac{1}{4}$ of 20 = ___

2 How much is shaded?
Choose $\frac{1}{4}$ or $\frac{1}{2}$.

a    **b**    **c**    **d**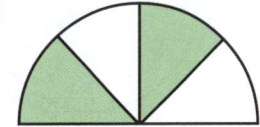

➡ *Workbook page 80*

More about time

What is the time?

> ### 💭 Think and share
>
> On a watch or a clock, the **hands** move around the **clock face**.
>
>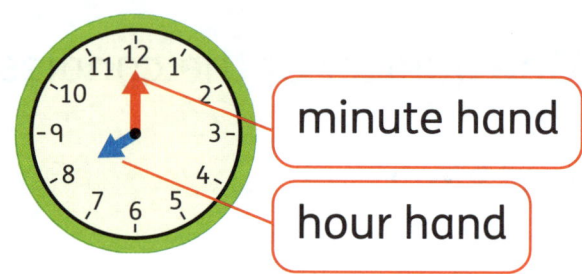
>
> This clock shows 8 o'clock.
>
> The **long hand** is the **minute hand**.
> The **short hand** is the **hour hand**.
>
> When the long hand points to the 12, the time is on the hour. We say **o'clock**.

1 Say the time on each clock.

a

b

c

d

➡ *Workbook page 81*

Different times of day

Zamir and Rani ate breakfast at 8 o'clock.

1 Look at the pictures.
They show what Zamir and Rani did at different times of the day.

a Say the time on each clock.

b Say what Zamir and Rani did at each time.

➡ *Workbook page 82*

Half past

In one hour, the minute hand makes a full turn.

In one hour, the hour hand moves to the next hour.

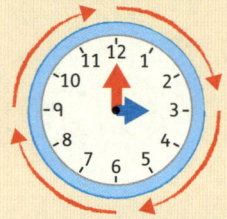 

It takes an hour to go from 3 o'clock to 4 o'clock.

One hour is 60 minutes.

Half an hour is 30 minutes.

The time is **half past** 3.

Halfway through the hour, the minute hand points to the 6, and the hour hand is halfway between the hours.

1 Say the time on each clock.

a

b

lesson continues ⬤

2   Say the time on each clock.

a

b

c

d

e

f

 **Problem solving**

3   The long hand points to 6 and the short hand is between 6 and 7.
What time is it?

4   I am thinking of a time on the hour (o'clock).
Both hands are pointing to even numbers.
One number is double the other one.
What time is it?

What doubles can you find on the clock?

Where will the long hand point?

My daily routine

Your daily routine is the way you do things every day.

This cartoon shows Tim's daily morning routine.

First, he wakes up.

Next, he gets dressed **before** his breakfast.

He brushes his teeth **after** his breakfast.

1 Write about your daily routine.

- What do you do first?
- What do you do next?
- What do you do after that?

2 Make sentences about the pictures.
Use before , after , first and next .

make a painting

wash my brush

lesson continues ▶

3 Make sentences about the pictures.
Use before , after , first and next .

play

pack away

4 What do you do first?
What do you do next?
What do you do after that?
Say the correct order.

let it cool

mix

cut the cake

bake the cake

5 What steps do you follow to wash your hands?
Say the steps in order.

➡ *Workbook page 83*

Use time words

The time is 6 o'clock.

Half an hour **later** will be half past 6.

An hour later will be 7 o'clock.

Half an hour **earlier** it was half past 5.

An hour earlier it was 5 o'clock.

1 What time will it be an hour later?

a

b

2 What time was it half an hour earlier?

a

b

3 Write sentences using these words.

yesterday   today   tomorrow   this morning

in the afternoon   in the evening

UNIT 20 Data

Use a Carroll diagram

Think and share

A **Carroll diagram** can help us sort.

Say how these objects were sorted.

	Fruit	Not fruit
Not yellow		
Yellow		

Where would these go in the diagram?

1 Use the same fruits and vegetables.
How would you sort them into this diagram?

	Vegetable	Not vegetable
Green		
Not green		

➡ Workbook page 84

Venn diagrams

A **Venn diagram** uses circles to sort things. The circles can overlap to show things that belong in both sets.

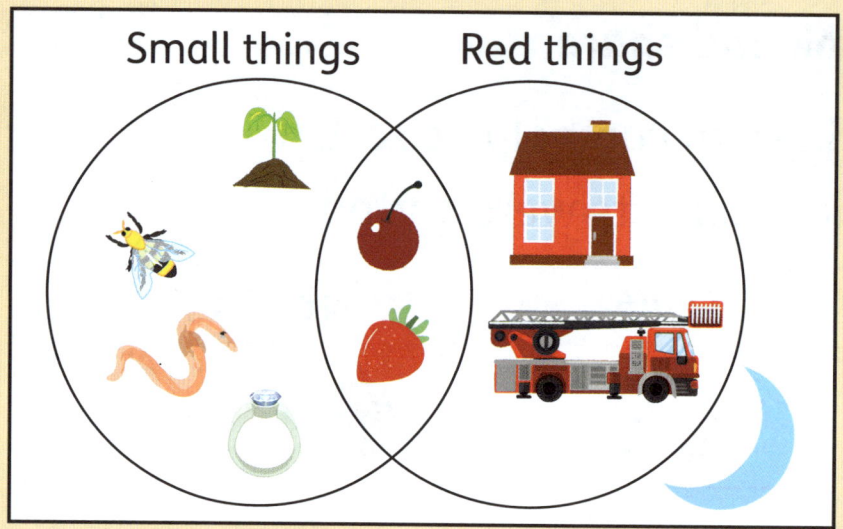

1 Say what you can see in the Venn diagram.

2 Where would these things go?

a   b   c   d

3 Where would the same objects go in this diagram?

	Small	Not small
Not living		
Living		

→ Workbook page 85

Pictograms

Trina drew this **pictogram** to show how many cars passed her house in one hour.

Number of cars that passed my house

Car colour	Number of cars
red	🚗🚗🚗
white	🚗🚗🚗🚗🚗🚗🚗
silver	🚗🚗🚗🚗🚗🚗
black	🚗🚗🚗
blue	🚗🚗🚗🚗🚗🚗

Key: 🚗 = 1 car

1   Answer the questions about Trina's pictogram.

    a   How many red cars passed?

    b   How many white cars passed?

    c   Which car colours had the same numbers of cars?

    d   Which colour was the most common?

> Most common means the same for the most people.

➡ *Workbook page 86*

More pictograms

Pictograms show **data** using pictures.

1 Ana collects data about favourite lunches.

Favourite lunches of Ana's friends

Lunch	Number of friends
sandwich	𝘩 𝘩 𝘩 𝘩 𝘩
noodles	𝘩 𝘩 𝘩 𝘩 𝘩 𝘩 𝘩 𝘩
rice	𝘩 𝘩 𝘩 𝘩

Key: 𝘩 = 1 friend

a What were the choices Ana gave her friends?

b How many people chose the sandwich?

c What was the most popular choice?

d Which lunch did the fewest people choose?

e How many people did Ana ask altogether?

➡ *Workbook page 86*

A cap pictogram

A pictogram must always have a key and a title.

1 Look at this pictogram and answer the questions.

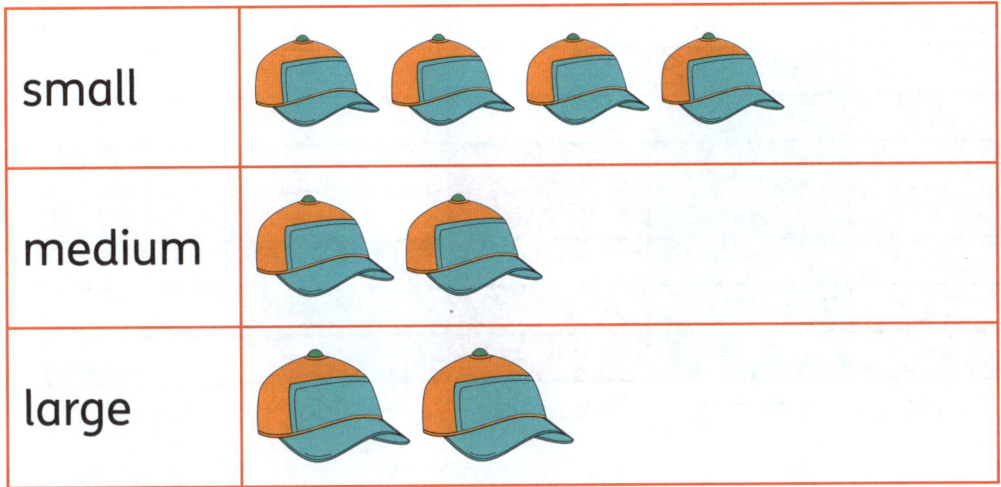

Key: = 1 cap

a What does the key tell us?

b How many large caps were there?

c Which two sizes had the same number of caps?

d The title is missing. What could the title be?

➡ *Workbook page 86*

Block diagrams

This is a **block diagram**. The numbers down the side give the **scale** of the diagram. They tell us how many people each block represents.

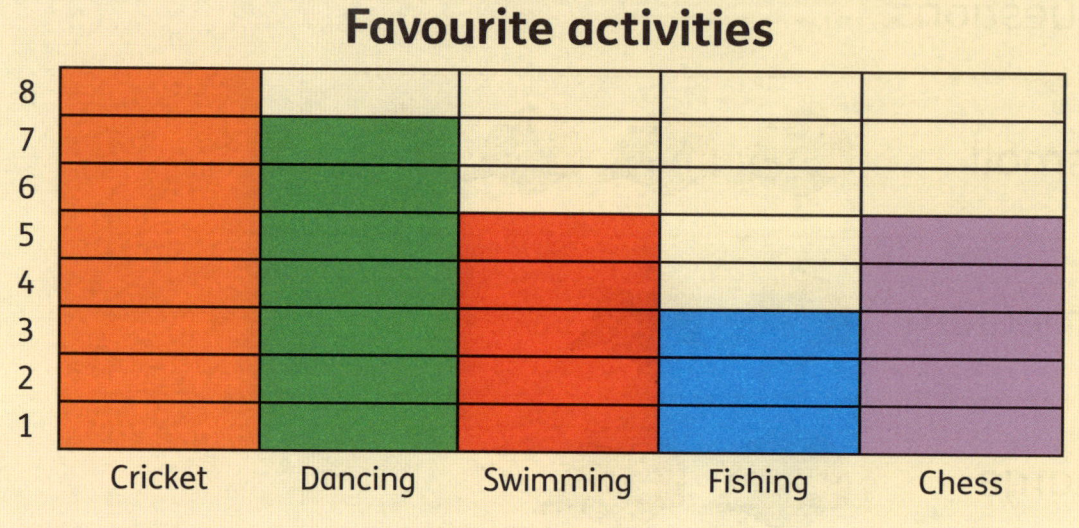

Favourite activities

1. **a** Which was the most popular activity?
   b Which activity did fewest people choose?
   c Which activities were equally popular?
   d How many more people chose dancing than chess?

 **Problem solving**

2. Find out the most popular after-school activities.
   - Pick five options for people to choose from.
   - Each person must only choose one favourite.
   - Find a way to collect the data.
   - Draw a block diagram to show your data.

➡ *Workbook page 87 and page 88*

Mixed practice 3

1 Write two sentences about this picture.
Use any four of these words or phrases:

on   in   under   in front of   behind   next to

2 An arrow starts pointing up.

  a Draw its position after a full turn.

  b Draw its position after a half turn.

3 Look at these coins from the UK.

1p   5p   10p   20p   50p   £1   £2

Draw a group of UK coins that have a
value of exactly 30p altogether.

4 Which of the following is not a property of
a rectangle?

- It has four straight sides.
- It has four corners.
- It has a curved outline.

5 What 2D shape is each person describing?

a
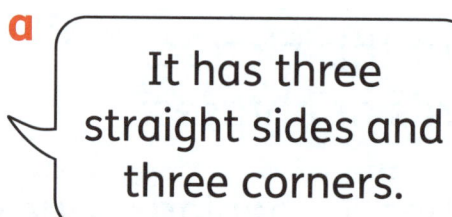
It has three straight sides and three corners.

b
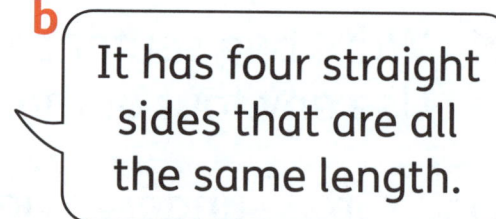
It has four straight sides that are all the same length.

6 How have these shapes been sorted?

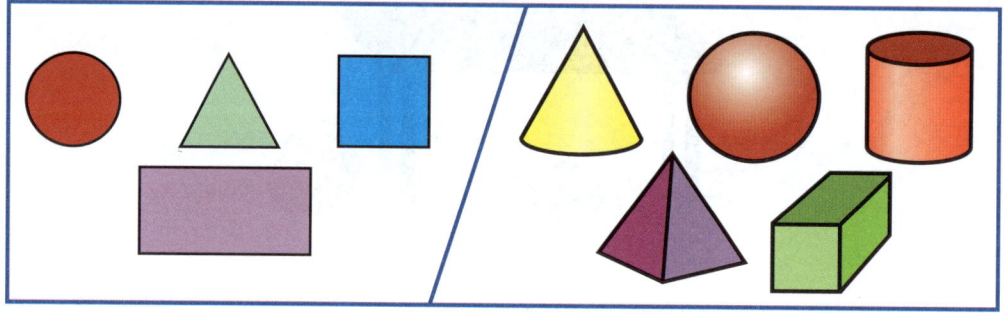

7 Which of these pictures does not show halves?

A   B   C   D

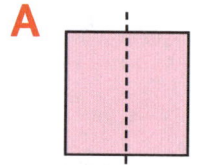 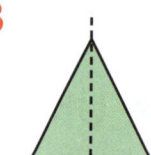 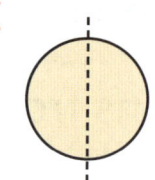 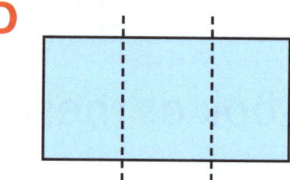

8 Why does this picture not show halves?

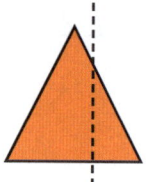

9 Which of these pictures shows quarters?

A   B   C

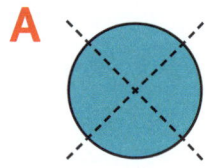 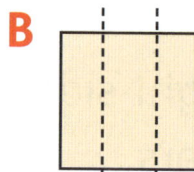 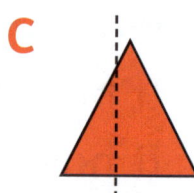

10 What is half of each amount?

a  half of 8

b  half of 12

11 Say the time on each clock.

a

b

12 Look at this pictogram.

How pupils in my class get to school

Method of travel	Number of pupils
🚲	⊙ ⊙ ⊙
🚗	⊙ ⊙ ⊙ ⊙ ⊙
🚌	⊙ ⊙ ⊙ ⊙ ⊙ ⊙ ⊙ ⊙
👟	⊙ ⊙ ⊙ ⊙ ⊙

Key: ⊙ = 1 pupil

a What does the pictogram show?

b Write three sentences about the pictogram.
Use the words most , fewest and same .

Glossary

+ – add or count on

– – take away or count back

= – makes or is equal to

2D shape – a form or outline with length and width

3D shape – a shape with length, width and height

A

Add – combine two or more numbers or amounts to make one number or amount called the total or sum; count on; plus; +

After – happening later than another event; following behind something

Array – objects or pictures put in equal rows and columns

B

Before – precedes or is in front of another time or object

Bigger – when something is larger in size than something else it is bigger; the opposite of smaller; the table is bigger than the pen

Block diagram – a chart made with blocks or rectangles

C

Capacity – the amount of liquid a container can hold

Carroll diagram – a type of table used to sort numbers or objects

Chart – a chart shows information or data

Circle – a 2D shape that is round

Clock face – the round surface with hours and minutes marked on it

Column(s) – objects or numbers set in a straight line from top to bottom

Cone – a 3D shape with a pointed end, a curved surface and a circular face

Corner(s) – the point where two sides of a shape meet

Cube – a 3D shape made of six square faces

Cuboid – a 3D shape made of six rectangular faces

Cylinder – a 3D shape made of two circular faces and one curved surface

D

Data – information that is collected about a topic

Date(s) – the day, month and year; 1 March 2000 is a date

Day(s) – one of the days of the week, for example, Monday or Thursday. Each day lasts 24 hours

Difference – you can find the difference between two amounts by subtracting or counting on; the difference between 4 and 6 is 2

Divide – to share things equally or group them into sets of the same size

Double – two times or twice as many

E

Earlier – happening before another time

Edge – the line on a 3D shape where two faces meet

Empty – holding nothing; the opposite of full

Equal(s) – has the same value or quantity

Estimate – to guess a rough answer using information you know

Even number – can be shared into two equal groups; even numbers end in 2, 4, 6, 8 or 0

F

Face – the 2D shape that makes one of the surfaces of a 3D shape

First – coming before all other objects or actions in a set

Full – holding as much as possible; the opposite of empty

Full turn – turning all the way around so you face the same thing as when you started

G

Greater – when an object has a greater mass than another object it is heavier; the opposite of lighter; the book has a greater mass than the pen; when an number is greater than another number it is larger; 16 is greater than 10

Greatest – the object with the greatest mass in a group of objects being compared; the number that is greatest in a group of numbers; 4, 7 and 16 – 16 is the greatest

H

Half – When you share things into two equal parts, each part is a half

Half past – 30 minutes past an hour; the minute hand points to the big 6, the hour hand is halfway between the hours

Half turn(s) – turning halfway around so you face the opposite direction to when you started

Hands (time) – the **hands** move around the clock face to tell us the time. The **long hand** is the **minute hand**. The **short hand** is the **hour hand**

Heavier – when an object has a greater mass than another object it is heavier; the opposite of lighter; the book is heavier than the pen, the pen is lighter than the book

Heaviest – the object with the greatest mass in a group of objects being compared; table, pencil, book – the table is the heaviest, the pencil is the lightest

Hour – one hour is 60 minutes

K

Key – a key tells you what a drawing or picture represents in a pictogram

L

Later – a word to compare times or to say that one time comes after another time or times

Length – how long something is

Less than – not as many as; not as much as; a small number or amount

Lighter – when an object has a smaller mass than another object it is lighter; the opposite of heavier; the pen is lighter than the book, the book is heavier than the pen

Lightest – the object with the smallest mass in a group of objects being compared; table, pencil, book – the pencil is the lightest, the pencil is the heaviest

Longer – when an object has a greater length than another object it is longer; the opposite of shorter; the table is longer than pen

M

Mass – how heavy something is; objects with greater masses are heavier than objects with small masses

Measure – we measure objects to find out how big or small they are

Minute – there are 60 seconds in one minute; 60 minutes make one hour

Month(s) – a month is about 4 weeks, or about 30 days; a year is 12 months; January is a month of the year

More than – a greater number or amount

Multiply – adding a number lots of times; 3 times 2 is 6

N

Narrower – when the length of something from left to right is smaller than the length of something else, we say it is narrower; the opposite of wider

Next – what happens after something else

Nought – another word for zero or 0

O

O'clock – the time when it is exactly on the hour and the long hand on a clock points to 12

Odd number – cannot be shared into two equal groups; odd numbers end in 1, 3, 5, 7 or 9

P

Pattern – objects, shapes, pictures or numbers that are organised to repeat

Pictogram – a type of chart that uses pictures to show information

Plus – another way to say add +

Properties – facts about a shape or number, for example a triangle has three sides; 2 is an even number

Pyramid – a 3D shape made of a flat base and three or more triangular faces faces that meet at a point

Q

Quarter – one of four equal parts of a whole; four quarters together make a whole; we write $\frac{1}{4}$ or one quarter

Quarter turn – a turn to face to the left or to the right of where you started; four quarter turns in the same direction make a full turn

R

Repeat adding – adding the same number many times; 2 + 2 + 2

Rectangle – a 2D shape with four straight sides

Row(s) – objects or numbers set in a straight line from left to right

S

Scale (for weighing) – an instrument used to measure mass

Scale (diagram) – a numbered line used to measure information on a chart

Second (time) – when a clock ticks, each tick is one second; there are 60 seconds in one minute

Second – 2nd; number 2 in order; one position or place after the first and before the third

Shorter – when an object has a smaller length than another object it is shorter; the opposite of longer or taller; the pen is shorter than the table

Side(s) – one of the lines of a 2D shape

Size – how big or small something is

Smaller – when something is littler than something else it is smaller; the opposite of bigger; the pen is smaller than the table

Smallest – the object or number that is smaller than all others in a group

Sphere – a round 3D shape; a ball is a sphere

Square – a 2D shape with four straight sides that are all the same length

T

Take away – to take a number or amount away from another number or amount; to count back; subtract; -

Taller – when an object has a greater length from top to bottom than another object it is taller; the opposite of shorter; the tree is taller than the car

Third – 3rd; number 3 in order

Three-quarter turn – three one-quarter turns in the same direction make a three-quarter turn

Total – the sum or whole amount; what you get when you add two or more numbers or amounts; the total of 2 add 2 is 4

Triangle – a 2D shape with three straight sides

V

Value (money) – how much something is worth. For example, a 5p coin and a 2p coin together have a value of seven pence

Venn diagram – uses circles used to sort objects, shapes and numbers into sets

W

Week – a week is seven days

Weighing – measuring mass using a scale

Whole – not divided into parts

Wider – when the length of something from left to right is greater than the length of something else, we say it is wider; the opposite of narrower

Y

Year – a year is 12 months; age is measured in years

Z

Zero – the word for the number 0